LONDON UNDERGROUND

1863 onwards (all lines and extensions)

First published in June 2014.
Reprinted in June 2015 and November 2016

A catalogue record for this book is available from the British Library.

ISBN 978 0 85733 369 8

Library of Congress control no. 2013956482

Published by Haynes Publishing,
Sparkford, Yeovil,
Somerset BA22 7JJ, UK.
Tel: 01963 440635
Int. tel: +44 1963 440635
Website: www.haynes.com

Haynes North America Inc.,
861 Lawrence Drive, Newbury Park,
California 91320, USA.

Printed in Malaysia.

Acknowledgements

A large number of people and organisations have helped me greatly in gathering information and images for this book, in particular:

Piers Connor; Tim Demuth; John Elson; Robert Excell (London Transport Museum Curator); Robin Gibson and Ian McKenzie of the London Underground Railway Society, who allowed me on several occasions access to the society's photographic archives, including the Bob Greenaway collection; Brian Hardy; Susan Harris (for giving me permission to reproduce her late husband's outstanding cutaway drawings); Colin Hayward and Robin Pinnock (Rothbury Publishing); Tony Howard (The Transport Design Consultancy); Daniel Imade (Arup Photo Library); Mark Ovenden; Andrew Parker; Paul Ross; Paul Rutter (DCA Design International); James Whiting (Capital Transport Publishing).

From London Underground/Transport for London:
David Waboso CBE (Capital Programmes Director); Mike Ashworth (LU Design and Heritage Manager); Andy Barr (LU Heritage Operations Manager); Daniel Hamblin (AIT Project Engineer); Tracey O'Brien (Senior Press Officer LU desk, TfL); Philip Shrapnell (Senior Rolling Stock Engineer); Alan Wilson (Project Manager, Tunnel Cleaning Train).

And last but not least my wife, Marilyn, for her support, patience and forbearance!

LONDON UNDERGROUND

1863 onwards (all lines and extensions)

Owner's Workshop Manual

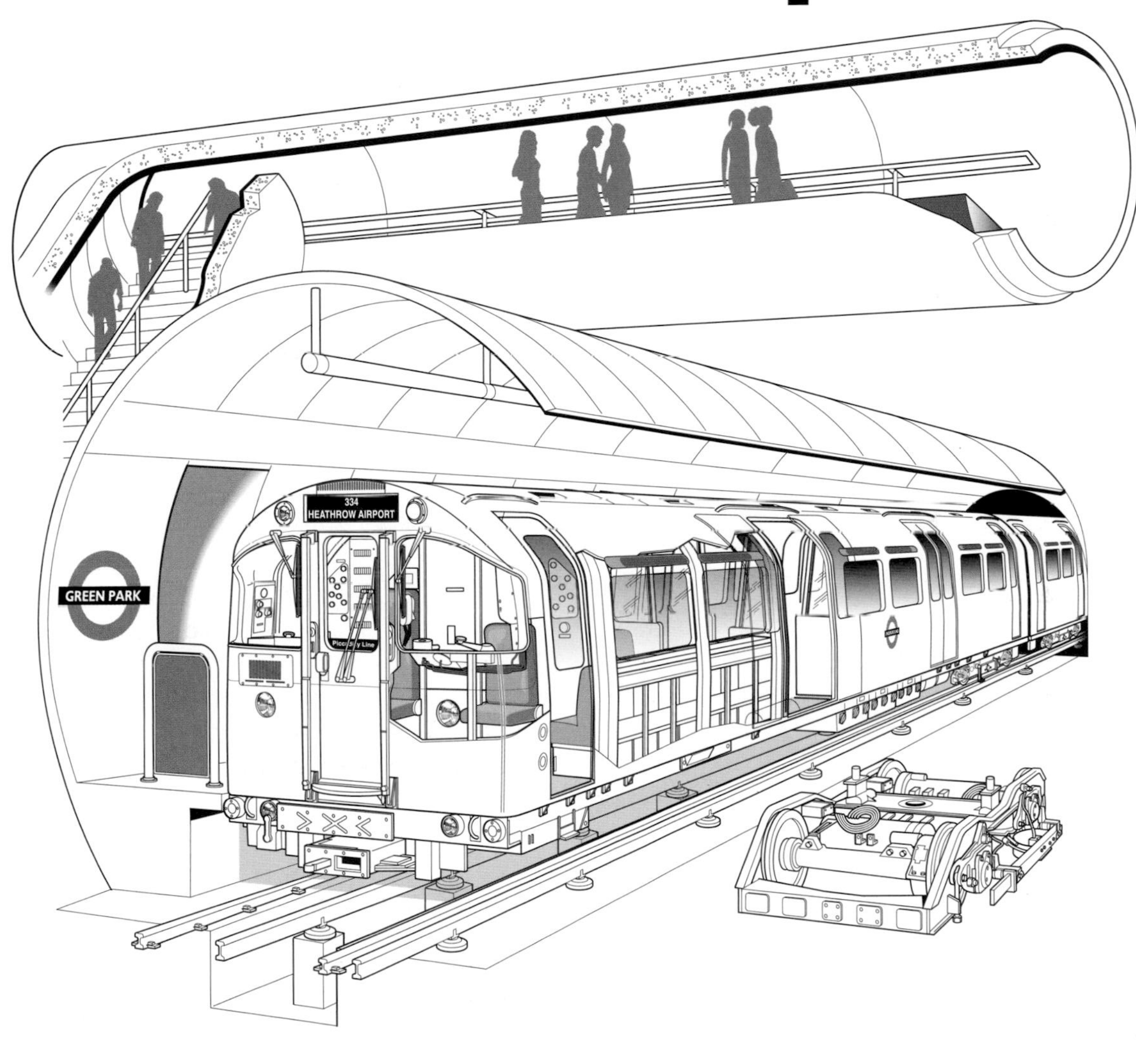

Designing, building and operating the world's oldest underground rail network

Paul Moss

NORTHFIELDS

NORTHFIELDS STATION
Local information

Contents

WOOD GREEN
TURNPIKE LANE
BRUCE CASTLE
TOTTENHAM HOTSPUR FOOTBALL CLUB
TOTTENHAM HALE
SEVEN SISTERS
MARKFIELD BEAM ENGINE & MUSEUM
FINSBURY PARK
MANOR HOUSE
ARSENAL
RESERVOIRS
CLISSOLD PARK
STROUD GREEN ROAD
HORNSEY ROAD
GREEN LANES
RIVER LEA
WALTHAMSTOW MARSHES
ARCHWAY
HOLLOWAY ROAD
HOLLOWAY ROAD
BLACKSTOCK RD.
HIGHBURY FIELDS
HIGHBURY & ISLINGTON
HACKNEY DOWNS
TUFNELL PARK
BRECKNOCK ROAD
CALEDONIAN ROAD
ALMEIDA THEATRE
CANONBURY ACADEMY
BALLS POND ROAD
KENTISH TOWN
HILLMARTON ROAD
BUSINESS DESIGN CENTRE
CALEDONIAN PARK
YORK WAY
ISLINGTON GREEN
ST. MARY'S
NEW NORTH ROAD
SHOREDITCH PARK
GRAND UNION CANAL
CAMDEN TOWN
KING'S CROSS STATION
CRAFTS COUNCIL
ANGEL
REGENTS CANAL
OLD STREET
CAMDEN LOCK
CAMDEN MARKET
GASWORKS
KING'S CROSS ST. PANCRAS
ST. PANCRAS STATION
PERCY CIRCUS
KING'S SQUARE GARDENS
PANCRAS ROAD
EVERSHOLT STREET
MORNINGTON CRESCENT
EUSTON STATION
EUSTON
NEW BRITISH LIBRARY
FARRINGDON
BARBICAN
CHARTERHOUSE
SMITHFIELD MARKET
LONDON ZOO
UNIVERSITY COLLEGE
RUSSELL SQ.
CORAM'S FIELDS
GRAYS INN RD.
CHANCERY LANE
NATIONAL POSTAL MUSEUM
MUSEUM OF LONDON
ST. PAUL'S CATHEDRAL
REGENT'S PARK
WARREN STREET
SENATE HOUSE
DICKENS' HOUSE
BRITISH MUSEUM
HOLBORN
SIR JOHN SOANE'S MUSEUM
LONDON SILVER VAULTS
PUBLIC RECORD OFFICE
ST. DUNSTAN
THE OLD BAILEY
OPEN AIR THEATRE
UNIVERSITY COLLEGE HOSPITAL
BANDSTAND
QUEEN MARY'S GARDEN
GREAT PORTLAND STREET
GOWER STREET
GOODGE STREET
BEDFORD SQUARE
CENTRE POINT
LINCOLN'S INN FIELDS
LONDON SCHOOL OF ECONOMICS & POLITICAL SCIENCE
ROYAL COURTS OF JUSTICE
DR. JOHNSON'S HOUSE
THAMESLINK
LUDGATE CIRCUS
BLACKFRIARS
REGENT'S COLLEGE
ROYAL ACADEMY OF MUSIC
POLLOCK'S TOY MUSEUM
TOTTENHAM COURT ROAD
KINGSWAY
THE OLD CURIOSITY SHOP
ROYAL OPERA HOUSE
ST. BRIDE
TEMPLE
PLANETARIUM
MADAME TUSSAUD'S
B.B.C RADIO
MIDDLESEX HOSPITAL
LEICESTER SQUARE
COVENT GARDEN
LONDON THEATRE MUSEUM
ALDWYCH
BOW ST.
SOMERSET HOUSE
COURTAULD INSTITUTE GALLERY
BLACKFRIARS BRIDGE
BAKER STREET
HARLEY STREET
WIMPOLE STREET
SOHO SQUARE
SOHO
EMPIRE
LONDON TRANSPORT MUSEUM
H.Q.S. WELLINGTON
H.M.S. PRESIDENT
MARYLEBONE
WALLACE COLLECTION
OXFORD CIRCUS
LIBERTY
CHINATOWN
SHAFTESBURY AV.
PLANET HOLLYWOOD
THE PIAZZA
THE STRAND
WATERLOO BRIDGE
EDGWARE ROAD
BAKER STREET
GLOUCESTER PLACE
BOND STREET
CAVENDISH SQUARE
HAMLEY'S
CARNABY STREET
TROCADERO
HALF PRICE TICKET BOOTH
NATIONAL PORTRAIT GALLERY
NELSON'S COLUMN
CHARING CROSS
CLEOPATRA'S NEEDLE
EMBANKMENT
NATIONAL THEATRE
MOMI
SELFRIDGES
OXFORD ST.
HANOVER SQUARE
NEW BOND STREET
SOTHEBY'S
BURBERRY'S
CAFE ROYAL
ROCK CIRCUS
THE DESIGN CENTRE
NATIONAL GALLERY
TRAFALGAR SQ.
ROYAL FESTIVAL HALL
HAYWARD GALLERY
MARBLE ARCH
EROS
PICCADILLY CIRCUS
HAYMARKET
DUKE OF YORK'S COLUMN
ADMIRALTY ARCH
MINISTRY OF DEFENCE
NEW MINISTRY OF DEFENCE
ARABIC CENTRE
SEYMOUR STREET
U.S EMBASSY
GROSVENOR SQUARE
BERKELEY SQUARE
ROYAL ACADEMY OF ARTS
FORTNUM & MASON
CHRISTIES
CARLTON HOUSE TERRACE
ICA
THE CITADEL
OLD ADMIRALTY
HORSE GUARDS PARADE
CABINET OFFICE
BIG BEN
JUBILEE GARDENS
SPEAKERS CORNER
PARK LANE
MAYFAIR
HARD ROCK CAFE
GREEN PARK
ST. JAMES'S PALACE
MARLBOROUGH HOUSE
DUCK ISLAND
DOWNING STREET
FOREIGN OFFICE
WELSH OFFICE
WESTMINSTER
LONDON AQUARIUM
HYDE PARK
WELLINGTON MUSEUM
PICCADILLY
LANCASTER HOUSE
CLARENCE HOUSE
THE MALL
ST. JAMES'S PARK
CABINET WAR ROOMS
PARLIAMENT SQUARE
WESTMINSTER BRIDGE
BANDSTAND
BIRD SANCTUARY
POLICE
CAFETERIA
CONSTITUTION ARCH
HOME OFFICE
WESTMINSTER ABBEY
HOUSES OF PARLIAMENT
PIER
THE SERPENTINE
HYDE PARK CORNER
BUCKINGHAM PALACE AND PALACE GARDENS
WELLINGTON BARRACKS
ST. JAMES'S PARK
NEW SCOTLAND YARD
DEAN'S YARD
BURGHERS OF CALAIS
THE VICTORIA TOWER GARDENS
HARVEY NICHOLS
KNIGHTSBRIDGE
GROSVENOR PLACE
BUCKINGHAM GATE
QUEEN'S GALLERY
PASSPORT OFFICE
DEPARTMENT OF TRANSPORT AND THE ENVIRONMENT
HYDE PARK BARRACKS
THE LIDO & CLUBHOUSE
SERPENTINE GALLERY
RESTAURANT
ROTTEN ROW
TOURIST INFORMATION CENTRE
WESTMINSTER CITY HALL
ROYAL MEWS
ARMY & NAVY
ST. JOHN'S CONCERT HALL
SMITH SQUARE
VICTORIA
HORSEFERRY ROAD
MARSHAM ST.
BELGRAVE SQUARE
ST. PETER'S
ROYAL HORTICULTURAL SOCIETY NEW HALL
ROYAL HORTICULTURAL SOCIETY OLD HALL
KENSINGTON GORE
HOLY TRINITY
BROMPTON ORATORY
HARRODS
SLOANE STREET
CHESHAM ST.
BELGRAVE PLACE
WESTMINSTER CATHEDRAL
PLAYING FIELDS
ROYAL ALBERT HALL
QUEEN'S TOWER
IMPERIAL COLLEGE
SCIENCE MUSEUM
PONT STREET
CHESTER SQUARE
EBURY STREET
VAUXHALL BRIDGE ROAD
ST. JAMES-THE-LESS
VICTORIA AND ALBERT MUSEUM
CADOGAN SQUARE
LENNOX GDNS
HOLY TRINITY
EATON SQUARE
ELIZABETH STREET
VICTORIA STATION
AIR/RAIL TERMINAL
ROYAL COLLEGE OF MUSIC
QUEEN'S GATE
BROMPTON ROAD
WALTON STREET
SLOANE SQUARE
ROYAL COURT THEATRE
VICTORIA COACH STATION
WARWICK SQUARE
ST. GEORGE'S DRIVE
ST. GABRIEL'S
PIMLICO
BELGRAVE ROAD
ST. SAVIOUR'S
SOUTH KENSINGTON
ST. MARY'S
PIMLICO RD.
WARWICK WAY
LUPUS STREET
CLAVERTON STREET
NATURAL HISTORY MUSEUM
BADEN POWELL
GLOUCESTER ROAD
MICHELIN HOUSE
DRAYCOTT AVENUE
SLOANE AVENUE
DUKE OF YORK'S HEADQUARTERS
LOWER SLOANE ST.
CHELSEA BARRACKS
WALPOLE STREET
HARRINGTON ROAD
ONSLOW SQUARE
ROYAL MARSDEN HOSPITAL
BURTON'S COURT
LISTER HOSPITAL
RANELAGH GARDENS
SUMNER PLACE
ROYAL BROMPTON HOSPITAL
THE CHELSEA FARMER'S MARKET
SMITH STREET
ANTIQUARIUS
ROYAL HOSPITAL
CHELSEA BRIDGE
EARL'S COURT
FULHAM ROAD
DOVEHOUSE ST.
HENRY J. BEANS
CHELSEA OLD TOWN HALL
NATIONAL ARMY MUSEUM
CHELSEA EMBANKMENT
CHELSEA PHYSIC GARDEN
ST. MARY-THE-BOLTONS
CRANLEY GARDENS
OAKLEY STREET
CARLYLE'S HOUSE
EARL'S COURT SQUARE
OLD BROMPTON ROAD
DRAYTON GDNS.
THE BOLTONS
BLUEBIRD GARAGE FASHION MARKET
KING'S ROAD
CHELSEA ANTIQUE MARKET
CADOGAN PIER
ALBERT BRIDGE
GILSTON ROAD
BEAUFORT STREET
CHELSEA OLD CHURCH
PEACE PAGODA
ED'S DINER
CROSBY HALL
REDCLIFFE GARDENS
PARK WALK
BATTERSEA BRIDGE
WEST BROMPTON
WESTMINSTER AND CHELSEA HOSPITAL
FINBOROUGH ROAD
LINDSEY HOUSE
BATTERSEA PARK
BROMPTON CEMETERY

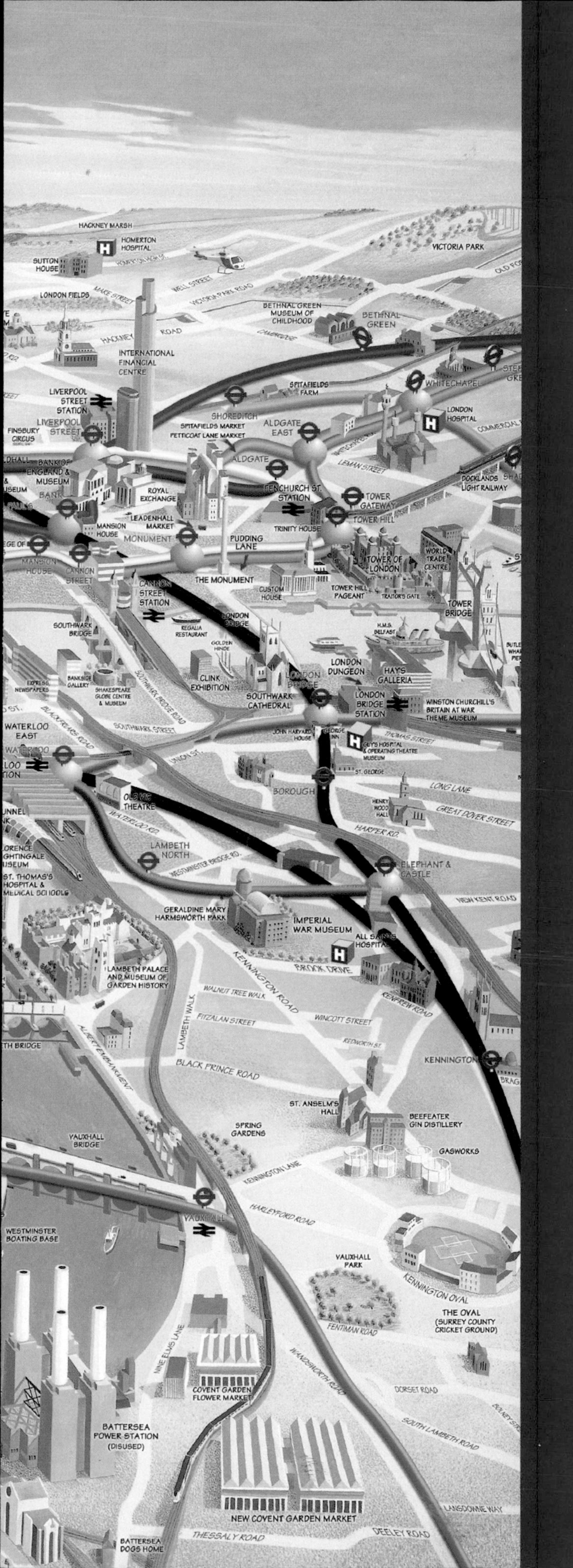

Introduction

London's underground railway system celebrated its 150th anniversary in 2013, making it the oldest – but no longer the largest – metro system in the world, and the history of its continual development makes for fascinating reading. It is a cornerstone of the day-to-day running and organisation of the nation's capital city and plays an incalculable rôle in making the country's economy function. Without it London could not perform properly, and would soon grind to a halt and cease to be a world-class financial centre.

In 2007 it carried its billionth annual passenger for the first time, and by 2011–12 this figure had increased to 1.171 billion passenger journeys per annum, a staggering increase of 64 million since the previous year! This demonstrates the unrelenting demands placed on its trains, stations and services, which continues to increase with every passing decade.

The network's routes cover 249 miles (402km), of which 55% are in tunnels, and there are 260 managed stations.

However, what is not generally known is that without American financial investment, design and engineering skills the system's crucial early expansion would in all probability never have taken place at all.

OPPOSITE **This 'Globalvision' map was prepared speculatively in the mid 1990s. Although trialled at one station, it was not taken up by London Underground for system use.**
(Copyright LondonTown.com)

DISTRIC
LUGGAGE
COMPARTMENT
10

PART ONE

The first period: 1863 to 1918

OPPOSITE **A District line 'B' stock train, with is American styling, prior to full electrification in 1905.**
(Bob Greenaway collection)

Chapter 1

Background to the first railway

150 years ago, a way had to be found to dramatically reduce the horse-drawn congestion on London's streets and to more effectively support and underpin the financial power and influence of the City of London. As a result of Great Britain's world-leading Industrial Revolution in the 18th century, London had by the 19th century become the largest and most prosperous commercial and industrial centre on earth, and the centre of the world's greatest empire.

London was then still a compact city, where the main-line railway termini such as Paddington and Euston had been built on its rural extremities, with no direct links through to the City. Passengers wishing to get to and from the City had three options:

- To use hansom cabs (if they were wealthy enough).
- To use horse-drawn omnibuses (even these were expensive, at a shilling fare per trip to travel from Paddington to the Bank).
- To walk, which the majority of people were consequently forced to do.

During the early 'boom' years of the dynamic growth of Britain's national railway system, the London & Greenwich – the first passenger-carrying railway in London – had by 1840 already carried nearly six million passengers to and from its terminus at London Bridge. This station was soon thereafter enlarged to create capacity for other railways coming in from the south, and all of these lines brought a daily influx of passengers who had only to cross the River Thames to enter the very heart of the City of London.

Other main-line services to feed into London were rapidly constructed, and today's main-line terminals were soon established. Trains from the north-west arrived at Euston (which opened in 1837), from the west at Paddington (1840), from the north at King's Cross (1852), from the east at Fenchurch Street (1841) and finally from the south-west at Waterloo (1848).

All of these termini were located some distance from the central business centres because of the high cost of constructing railways through built-up areas as a result of the unavoidable and expensive demolition of property along the chosen route. Consequently of the main termini only one, Fenchurch Street, had been built to directly access the City of London's 'square mile'. It was this lack of direct rail services into the City that was to provide the main impetus to construct an underground railway beneath Victorian London.

With the ever-increasing volume of horse-drawn traffic on the streets, much of it being caused by the railways themselves, congestion was rapidly becoming a nightmare. It was claimed that it took longer to get across the centre of the metropolis from London Bridge to Paddington than it did to travel up to London Bridge from Brighton! It is also significant that by 1850 the population of Greater London had increased to 2.5 million from fewer than a million just 50 years earlier. As a result, nearly a quarter of a million people entered the City every day to work.

Charles Pearson makes his entrance

A Parliamentary Select Committee report on Metropolitan Communications that was published in 1855 recommended that the main-line stations in London should be linked by an underground railway. This was, however, not a new idea, as it had been first suggested back in the 1830s. One of its first supporters was the visionary and altruistic City of London Corporation's solicitor, Charles Pearson, who seized on the economic and social advantages that would be created by linking the main-line railways through to the City. His pet project was for an underground railway that would join up with the main-line termini, because linking them via a surface railway using a series of viaducts where necessary would be far too costly to be realised.

A significant railway development around the periphery of the capital had been the opening of the North London Railway (as it came to be called), which boasts an unbroken record of service to the present day (now as part of the recently reconstituted 'London Overground' network). This line started to run westwards from Richmond, first using the main line to West London Junction near Kensal Green and then, from 1860, along a new line through Kentish Town (Gospel Oak), Hampstead Heath, Finchley Road and Kilburn.

This venture's early success paved the way for the possibility of another line running further into Central London under established residential areas. The City Corporation wanted to remove its insanitary cattle market and the totally unacceptable slum conditions of its dwellings and the back-street manufacturing industries within the area of the Fleet Valley.

Moreover, Pearson realised that railways could run trains with a very cheap fare structure to enable working people to live further afield from their place of work, which had two major advantages: they would enjoy a healthier living environment in their leisure hours, with the additional bonus of paying lower rents for their accommodation.

Pearson had proposed a railway that would run under a new road down the Fleet Valley from King's Cross to a large goods and passenger terminus at Farringdon Street, but this scheme could not attract the required financial support. However, it sparked off the notion amongst financiers that a railway connecting the Fleet Valley, built under the busy Euston and Marylebone roads to Paddington, would be a very sound business proposition and deserved their full backing. The plan was to join the Great Western Railway there and the Great Northern Railway at King's Cross. Their financial support would be crucial, since the prize would enable them to run through trains into the City.

LEFT Charles Pearson. *(London Transport Museum)*

The money is found

It took the whole of the 1850s to convince the City Corporation that its financial support was not only in its own environmental interest but also essential to the scheme's eventual success. Investors were sceptical about persuading people to travel below ground through gloomy and smoky tunnels, and that such a railway would therefore never be profitable. However, the promise that the GWR would give its financial support in return for a junction with its main line at Paddington – which would enable through running to gain access to the City – was the catalyst. A final revised scheme for an underground line from Paddington to Farringdon Street via King's Cross gained its Act of Parliament in 1854, but Charles Pearson struggled on for a further five years to gather the estimated £1,000,000 of finance needed to build it. He persuaded the City Corporation to contribute £200,000, the Great Western then put in £185,000, and the balance of £615,000 was provided by private investors by 1859. Construction of the Metropolitan Railway, as it would be named, could finally begin.

RIGHT Digging under the Marylebone Road. *(London Transport Museum)*

The route is planned and built

Wherever possible the railway was to follow the road above, in order to avoid paying the huge costs incurred through the demolition of property and subsequent compensation. For example, the section from the Fleet Valley under Mount Pleasant was tunnelled, but the rest, under the Euston and Marylebone roads – then as now a major traffic highway – was built using the 'cut and cover' method, in which the road was dug up and a trench created about 15ft deep and 30ft wide. The side walls were lined with bricks, the top lengths of which were linked to an elliptical brick arch that formed the roof. The road surface was then restored.

In places where there was insufficient topsoil available to build such an arch, cast or wrought iron girders were used instead to complete the rectangular supporting 'box' structure. To save damaging the foundations of property topside, the trench was dug along the line of the above-mentioned roads, but the unseen drawback of this was having to negotiate a path through the established below-ground services, such as water, gas and sewage pipes. Just as in similar constructions today, these often had to be diverted, which soon forced the construction costs above the original estimates.

As might be expected, the building of the first stretch of line was not without significant social costs. Although the official figure for the people displaced by the building works was just 307, other estimates put the number as high as 12,000 after 1,000 homes had been swept away, with scant amounts of compensation on offer.

Steam traction for the first forty years

Because steam traction was the only technology available to haul carriages, blowholes were positioned at points in the middle of the roads to allow fumes and steam to escape. The section of the route between King's Cross and Farringdon Street was built partly in an open cutting, but included the only true tunnel on the line. This ran for nearly half a mile under Clerkenwell and was, at its deepest, almost 60ft below street level.

Because the GWR had contributed to the necessary financial backing required to get the railway built in the first place, Brunel's 'broad gauge' trains with their 7ft ¼in track width had to be accommodated from the outset. This dimension was unique to that railway, but the almost universal (then and now) track width of 4ft 8½in (established by railway pioneer George Stephenson in 1826) was also incorporated by

LEFT The River Fleet is breached. *(London Transport Museum)*

means of an inset third rail. This soon proved to be a very fortuitous decision.

When completed, the first section of the Metropolitan Railway had cost £1.3 million to build, a 30% increase over the original cost estimate – a state of affairs that remains all too familiar today! One unseen cost was due to the River Fleet, which had been diverted into a brick sewer alongside the railway but unfortunately burst through the retaining walls near Farringdon Street and flooded the railway lines to a depth of 10ft, right back to King's Cross! Luckily no one was injured, and it only took six weeks to clear the vast amounts of debris and 'make good' once more. After this delay and another one caused by initiating required changes to the block signalling apparatus (designed by the GWR), the first urban underground railway finally opened for business on the 10 January 1863.

A huge army of navvies worked on the construction of London's underground railways throughout the 1860s, and tough, back-breaking work it was too, often working in the harshest, most primitive conditions. Their fine workmanship was nevertheless in the tradition of the very best standards achieved by Victorian engineering, and the quality of the brickwork in these early stations with their brick retaining walls is still to be enjoyed and marvelled at today.

The vital role of all-embracing engineer in charge of the whole project was undertaken by John (later Sir John) Fowler, who was paid the astronomical figure of £152,000 (about £11.5 million in today's money). Although he had, of course, to hand over a good proportion of this amount to pay all his colleagues and assistants, it was nevertheless a very handsome amount of money! But he certainly delivered the goods, by taking what was at the outset a completely untried – and for many a somewhat far-fetched

LEFT Sir John Fowler. *(London Transport Museum)*

ABOVE Early workmen's train. *(London Transport Museum)*

BELOW Daniel Gooch's broad gauge locomotive *Wasp* at Baker Street. Painting is by C. Hamilton Ellis and shows Sir John Fowler and Sir Benjamin Baker (responsible for the tunnelling). *(London Transport Museum)*

– concept to a successful conclusion. He would remain the consulting engineer on some of the other Underground lines up to the turn of the century, including the first electric 'tube' railways.

The new railway was very well patronised right from the start. Thousands of people queued for tickets on the first day, with nearly £850 taken in receipts, and in its first six months of operation it was carrying an average of 26,500 passengers a day. The first-class single fare was 6d (about 2.5p), half the price of horse-drawn omnibus fares for the same journey, and the second- and third-class fares were 4d and 3d.

In its first year of operation 1.8 million people used the line, a daily average of 32,300, and as a result the Metropolitan's receipts were burgeoning, with a profit of £102,000 being achieved in its first year. Generous dividends of 6.25% were paid to shareholders in 1864, a much better return than most other railway companies gave.

In May 1864 the Metropolitan made £720 per mile per week, compared with the £80 of the London, Chatham and Dover Railway, the next best performer. In retrospect, this proved to be the line's honeymoon period, for such generous payouts would never be made again.

From the spring of 1864 the Metropolitan finally made one of Charles Pearson's chief objectives a reality, by offering cheap early morning fares of 3d return (later reduced to 2d) – the first railway to do so. Two trains started running in each direction before 6:00am, their passengers having the right to return by any train. This clever marketing ploy greatly increased the attractiveness of the line, and these 'workers' trains would eventually constitute 70% of the passengers carried in its early years. But sadly Pearson died just four months before the first of these trains ran, and the man who we might call the 'Godfather' of the Underground therefore never lived to see his dream become a reality. Pearson was nevertheless the first of a select group of immensely talented and visionary men who would shape the fortunes of the Underground network over the coming decades.

The early locomotives

In the beginning all of the rolling stock came from the GWR, still running on their broad gauge track system. Forty-five eight-wheeled compartment coaches were supplied which were illuminated by coal gas lamps fuelled from inflatable bags on the carriage roofs. Daniel Gooch, the Great Western's chief locomotive engineer, designed and built 22 coke-burning engines especially for the locomotives' unique operating conditions. They were successfully fitted with a condensing system that enabled steam to be channelled back into cold-water tanks instead of blowing out into the running tunnels.

Within a few weeks of opening,

disagreements arose between the Metropolitan's directors and those of the Great Western over financial and operational issues. In principle, the Metropolitan required a more frequent service than the GWR was prepared to offer. Petty squabbles escalated to the point where the GWR threatened to withdraw all of its rolling stock within the next few weeks, and selling it to the Metropolitan was not an available option.

The Great Northern came to their assistance and, via a new link line at King's Cross, was able to provide standard gauge rolling stock at short notice on a temporary basis. The Metropolitan had to move fast and initially ordered 18 standard gauge 4-4-0T locomotives from Beyer Peacock in Manchester, but eventually the order would increase to 66. These engines incorporated a similar condensing system to Gooch's design, and deliveries commenced in 1864. They performed with great safety and reliability for the next 40 years, and the sole survivor can today be seen at the London Transport Museum in Covent Garden, London.

This was an astounding record given the operational conditions; it proved that the designer and the manufacturer got it right first time. New coaches built on the GWR model were also procured from the Ashbury Railway Carriage and Iron Company, also based in Manchester. An innovation that did much to dispel any lingering doubts and fears about travelling through dark, smoky tunnels was the introduction of very bright gas lamps in these carriages. The gas was contained in long rubber bags clad in wooden boxes that were mounted on the roofs and then piped through to the twin burners of the lamps.

Safety had been the essential building block of this remarkable success story. The railway's simple but effective signalling system meant that throughout its first 44 years of steam-hauled operation it did not experience a single accident that resulted in the death of a passenger. This was a staggering achievement given the intensity of service, the use of steam engines (with the continual problem of operating in a smoke-filled and steam-laden environment) and wooden carriages, and the high number of passengers carried.

ABOVE The surviving Beyer Peacock locomotive. *(London Transport Museum)*

BELOW Restored 1892 'Jubilee' coach No 353 at Baker Street, the oldest operational Underground carriage in existence. *(Andy Barr)*

Chapter 2

Rapid expansion

Meanwhile, buoyed up by the underground railway's initial success, expansion of the original route proceeded at a heady pace. The Metropolitan's main *raison d'être* was to service the needs of the City, which was a magnet for employment throughout the region. The directors had planned for growth right from the beginning, and had obtained authority to construct two extra tracks between King's Cross and Farringdon, called the 'City Widened Lines', which very soon provided an important connection with the Midland Railway at St Pancras.

This was the springboard for future development, and additional Parliamentary powers were soon gained to enter into the City itself via a new terminus at Moorgate Street that opened at the end of 1865. But before this, in June 1864, an extension westwards used the GWR main line for about a mile from Paddington (running, of course, on dual-gauge tracks) on to a new railway (the Hammersmith & City, formerly independent but later co-owned by the Metropolitan and the GWR). This ran through the then pastoral fields south of Portobello and the Notting Barn Farms to a terminus near the north side of Hammersmith Broadway, via new stations at Latimer Road and Shepherd's Bush. These new stations stimulated demand for speculators to construct new properties around them, thereby setting a pattern that would

BELOW The terminus at Farringdon. *(London Underground Railway Society/LURS)*

ABOVE **Cutting through housing at Praed Street.** *(London Transport Museum)*

ABOVE RIGHT **Paddington Station today with original open area to get rid of smoke and steam.** *(Author's collection)*

BELOW **Advancing towards Notting Hill Gate.** *(London Transport Museum)*

ABOVE **Original drawing of Notting Hill Gate station.** *(London Transport Museum)*

LEFT **The station today.** *(Author's collection)*

ABOVE St John's Wood station, 1868. *(London Transport Museum)*

be repeated for the next 150 years, with the Underground's arrival making many developers their fortune. An additional new branch, running out to St John's Wood and Swiss Cottage, was operational by1868, and again the area was rapidly developed, with attractive 'villa' style homes for the middle classes.

From the day of its opening the success of the Metropolitan knew no bounds, and Parliament soon appointed another Select Committee of the House of Lords to recommend the best additional routes which would further improve transport services in the capital. The committee wanted to link all of the other main-line termini built or to be built north of the Thames, and considered that the best option was to extend the Metropolitan in both directions and form an 'inner circuit' to serve them. The Metropolitan obtained powers to continue its line from Moorgate Street, then around the City on one route while another headed out from Paddington via Bayswater, Notting Hill Gate and High Street Kensington to Brompton (later to become South Kensington). Another company was subsequently formed, the Metropolitan District Railway, which was created to build the section of the Inner Circle between Tower Hill and South Kensington. The two railways were closely connected, as there were four Metropolitan directors on the District's board, and it was expected that they would merge on completion of the Inner Circle.

Creating lines through these areas of London was at once more expensive and complicated than the initial 'cut and cover' route because the new tunnels were driven alongside and through extensive property development, which as before entailed expensive compulsory purchase, demolition and compensation costs. The original plan for the Metropolitan's western section of line was to build it through the open lands of Kensington Gardens and Hyde Park, but understandably this met with huge opposition. And so the alternative route that we have today, from Paddington to Gloucester Road, was opened on 1 October 1868, and pushed onwards to Brompton (South Kensington) almost three months later.

RIGHT Pushing towards South Kensington. *(London Transport Museum)*

LEFT Digging the Embankment for the District line. *(London Transport Museum)*

The District line emerges as a separate company

The building of an embankment along the Thames foreshore which had been started in 1862 (named the Victoria Embankment) was a very timely development as far as the Metropolitan District was concerned, because it allowed them to marry these massive civil engineering works with a new railway. The plan was that once completed, all of this installation would all be covered over by a wide roadway, enabling the intense traffic bottle-neck on the parallel Strand and Fleet Street routes to be relieved (which it still does very effectively today).

A shortage of cash delayed the start on this section but it eventually opened for service between Westminster and Blackfriars in May 1870. But even before this landmark event that should have ensured a close co-operation between both railways, there were already signs of a rift between them that was to later escalate. The independent directors of the District had become increasingly displeased with the operation of their services by the Metropolitan. As a result the Metropolitan was given the required 18 months' notice to end their operating arrangement, and their four directors on the District's board resigned. The District's riposte was to seek independence where they could and appoint their own managing director.

Striking out on their own, the District built their own repair works and sheds at West Brompton and started to buy their own locomotives. These were a repeat order from Beyer Peacock of the Metropolitan's own 4-4-0Ts, but painted in green with 'District Railway' on the tank sides. The initial order was for 24, but over the next 15 years their fleet would increase to 54. They began operating their own services in July 1871, when a short extension from Blackfriars to Mansion House became their terminus. Throughout the 1880s rivalry between the two companies was fierce, as they vied with each other via competitive fare structures and timetabling.

However, neither company showed much eagerness to complete the Inner Circle, as both were in a shaky position financially. Overground extensions into the leafy undeveloped suburbs seemed a more likely way to generate profits than costly cut-and-cover constructions underneath the City, so these were built rapidly and in great profusion. The District built its own link to Hammersmith in 1874, providing a more direct route into the City than the Metropolitan's service via Paddington. (This explains why today there are two Underground Hammersmith stations just a couple hundred of yards apart.) They obtained running powers over the London & South Western Railway to push through to Richmond by 1877 to make a connection with the previously mentioned 'North London line.'

A short line via a spur from an LSWR junction at Turnham Green to Ealing Broadway followed in 1879. The West Brompton terminus was pushed through to Fulham Broadway and to Putney Bridge, opening in 1880. Nine years later a new bridge over the Thames took the line to Wimbledon, running from East Putney on LSWR tracks over which the District had secured running powers. This situation lasted until as recently as 1994, when British Rail (who inherited the stretch of line from the Southern Railway) finally surrendered this group of stations to London Underground following privatisation.

All of these extensions were driven through mostly virgin farmland and set a pattern that continues to this day in that property is soon developed close to new stations, resulting in ever-increasing values. By 1880 much of today's District line routes that bring in tens of thousands to their places of work in central London had been already established.

The intended Circle line remained open-ended for some time, for apart from linking Moorgate Street to Aldgate in 1876 the cost of the City part of the scheme was too expensive, and the required finance wasn't available to fund its completion for another eight years. However, in September 1884 the circle was finally closed and trains could run round its whole perimeter.

Fierce rivalry

Understandably, the pioneering underground line was not going to play second fiddle to the District when it came to expansion. An extension from Moorgate to link the important main-line Liverpool Street station was opened in mid-1875 with its own station, called Bishopsgate. Swiss Cottage was extended first to West Hampstead and Willesden Green in 1879 and then right out to Harrow in 1880. The speed of progress in constructing these rail extensions remains an astonishing achievement, but of course, they were being laid through open countryside.

The intense rivalry between the Metropolitan and the District lines was intensified by personal hostility between their respective chairmen, the District's James Staats Forbes and his Metropolitan counterpart Sir Edward Watkin, whose animosity towards one another would gradually intensify into outright hatred (and was a key reason why the Inner Circle took so long to complete). The District line remained true to its *raison d'être* as an urban railway with a number of branch extensions, but the Metropolitan had much greater aspirations. However, in order to grow it had to reach much further afield, because any extension through the north-west suburbs was constrained by the tracks of other railways and there was consequently very little scope to build significant branches that made any business sense.

By 1888 the Metropolitan had already expanded well beyond Harrow. Pinner was reached in 1885, Rickmansworth in 1887 and Chesham by the middle of 1889. Yet Watkin wanted his railway to be a far larger concern.

RIGHT Painting showing Metropolitan and District line trains operating the Circle line, meeting at Baker Street. *(Copyright Robin Pinnock/ Rothbury Publishing)*

FAR LEFT **James Staats Forbes.** *(London Transport Museum)*

LEFT **Sir Edward Watkin.** *(London Transport Museum)*

His vision was for it to become the pivotal cog of a main-line network that would run from Manchester via London to Dover, then through a channel tunnel to Paris and the rest of Europe! Nor was this deeply impressive vision just a pipedream, for he was also the chairman of the Manchester, Sheffield and Lincolnshire Railway, which was preparing a new trunk route to London from the north-west. He told his Metropolitan shareholders in 1888 that 'I do not intend to be satisfied, if I live a few more years, without seeing our own railway become the grand terminus for a new railway system throughout England!'

But in order to fulfil his ambition to operate a main-line steam railway, he had

BELOW **Speeding to Rickmansworth in the 1880s.** *(London Transport Museum)*

LEFT Ashbury-built bogie coaches from this period. *(LURS collection)*

to spend a significant amount of money on new equipment. The short four-wheeled carriages that were introduced from 1887 onwards were not up to the demands of the extended lines, so from 1898 onwards these were gradually replaced by bogie stock, which featured electric lighting, steam heating and upholstered seating, and had new steam locomotives to haul them. The relationship between the rival companies started to improve after the resignation of Sir Edward Watkin following a stroke, but as the 19th century was drawing to a close technological developments were starting to create groundbreaking new possibilities for underground travel.

ABOVE 1898 Metropolitan locomotive No 1 in restored condition. *(Bob Greenaway collection)*

BELOW The same engine decorated with flags and bunting when it hauled the first passenger train to Uxbridge in July 1904. *(Bob Greenaway collection)*

The development of tunnelling techniques

The disadvantages of cut-and-cover tunnel construction had been learnt the hard way. It had contributed to the long delay in completing the Inner Circle, and was no longer a viable tunnelling option for private developers in the Central London area, where there were not enough straight roads to tear up, as had previously been the case, and the first 12ft or so below the surface was by now full of sewer systems and gas, electricity and water conduits. Furthermore neither the Metropolitan nor the District had been able to demonstrate any real financial payback due to the huge constructional costs they had to bear: tearing up the capital's roads in such a drastic manner, as well as problems in raising the necessary finance, made it difficult to guarantee the success of any future grand schemes of this type. And yet there remained a clear need for some form of rapid transit in Central London, particularly in the West End.

The technological breakthrough that allows the next part of our story to be told was prefaced by an invention by Marc Isambard Brunel – the French-born father of arguably Britain's greatest engineer of the 19th century, Isambard Kingdom Brunel – who was involved

ABOVE Digging the Thames Tunnel. *(London Transport Museum)*

in the earlier attempts at deep-level tunnelling beneath London that predated the Metropolitan Railway's construction by several decades. As early as 1818 he had patented a technique of tunnelling through London's green clay using a protective shield. The first application of this method was in the construction of the first successful tunnel under the Thames in 1825, and a development of the same concept would make it possible to delve deep tunnels for trains in the future.

Brunel's shield was rectangular in shape and was made up of 12 independent cast iron frames, each of them further divided into compartments to form a total of 36 open-ended cages. The frames were sunk to the bottom of a vertical shaft that was just over 60ft deep. They were then inched forward into the earth with miners inside them, who hacked away the earth with picks and shovels, the soil being removed in barrows as they advanced. The shield was then moved forward by a series of jacks, and bricklayers following behind the miners lined the new tunnel walls and roof with bricks to prevent the earth collapsing. Nevertheless, there were frequent floods during the course of construction, one of which killed six men and almost drowned Brunel himself.

The money put up by investors soon ran out, but the government bailed out the project, and the tunnel finally opened in March 1843, nearly 20 years later and at over twice the projected cost. The Thames tunnel, designed for horse-drawn carriages and pedestrians, was a landmark engineering achievement, as it was the first to be built below a body of water, but it was also a financial disaster. From favourable beginnings as a tourist attraction, by the 1860s it had declined to

ABOVE Train running through the tunnel in 1870. *(London Transport Museum)*

become a haunt of drunks and thieves. Finally it was sold to the East London Railway line in 1870, as a running tunnel. However, because its connections to the rest of the network were poor the East London line soon suffered financial difficulties and was snapped up by Watkin in 1878. Although services improved under his management, it would remain the Metropolitan's poor relation for over a hundred years, with mediocre levels of ridership. However, its time would eventually come as part of the excellent London Overground system in the 21st century.

That Brunel's major engineering enterprise has endured the test of time so successfully is a massive endorsement of his amazing vision.

RIGHT Sir William Barlow. *(London Transport Museum)*

The Tower Subway – London's first tube railway

The Tower Subway is the technological link between Brunel's Thames Tunnel and the first electrified tube railway. It used a new type of shield excavating system patented by Peter William Barlow in 1864. This was a far simpler method than Brunel's, with a much smaller bore diameter (6ft 8in), but its superior tunnelling technology enabled the Tower Subway to be completed in only five months. The 'Barlow Shield' therefore constituted the next major step towards the definitive method of boring tube lines through London's green and grey clay for the next 100 years. In essence it created cylindrical rather than rectangular tunnels, and a lining of curved cast iron segments replaced Brunel's brickwork. At the cutting edge of the shield one or two workers enclosed in a cavity excavated the clay from the tunnel mouth with picks and shovels as before. At the rear, other workers also within the shield's protection bolted the iron segments together to form rings, and at regular intervals the shield was forced forward by screw jacks set against these lining rings.

The clever and innovative solution to maximise the safety and integrity of the construction and the workers' lives was to pump quick-setting lime grout into the gap left by the shield between the tunnel rings and the surrounding soil. This prevented the soil from settling and causing subsidence.

The Subway was located downriver from London Bridge, and a single small carriage with a carrying capacity of 12 was hauled through the tunnel by a cable powered by a fixed steam engine. Two classes of fares were

offered, but 'first' only guaranteed that you were first in the queue for the steam-driven lifts. However, the project was soon regarded as a commercial failure, because the revenue taken could not support the operating costs. These were compounded by mechanical failures, and this short line was consequently declared bankrupt after only three months in operation.

The train and all of the operating equipment, including the lines and lifts, were removed, and the tunnel became a pedestrian walkway with a toll payment of one halfpenny. This was certainly a sound commercial venture because one million people a year went on to use it, and it only went out of business in 1894 after the Tower Bridge was opened.

ABOVE Building the Tower Subway, which opened in 1870. *(London Transport Museum)*

LEFT Passengers boarding the cable car. *(London Transport Museum)*

Chapter 3

Electrification

Electric traction develops

Although no further deep tube railways were to be built for a further 20 years, the system of boring tube tunnels deep beneath London was now proven. However, new traction technology needed to become available before advantage could be taken of this. The solution that initiated the next major stage in the development of the London Underground resulted from pioneering work carried out in Germany, where Werner von Siemens had demonstrated the capabilities of a small electric locomotive designed by him in 1879, which hauled a train of visitors around the grounds of an international exhibition in Berlin. In 1883 this new technology was used on England's first public electric railway – the Volks Electric Railway on Brighton's seafront that's still running today.

In 1884 Parliament authorised the construction of an underground railway to be called the City of London and Southwark Subway. Two tunnels just over 10ft in diameter were to be bored under the Thames at a depth of 60ft, to run between the Elephant and Castle and King William Street, a distance of approximately 1.5 miles. One of the line's leading advocates was James Henry Greathead, a South African engineer, who had overseen the building of Barlow's tunnel subway and later became the resident engineer for the District's Hammersmith extension. He was appointed Civil Engineer in charge of constructing the tunnels. He improved Barlow's horizontal cylindrical tunnelling shield to the point that it became the optimal design for all subsequent tunnelling schemes in London. Having seen the San Francisco cable railway in action, he was initially wedded to the same operating system where the cars would grip the moving cable en route to stations and then 'ungrip' on arrival. However, the contractor for the cable system went bankrupt during construction and additional doubts were soon raised, particularly as further powers were obtained in 1887 to double the line's original length to three miles.

BELOW Surviving locomotive and 'padded cell' carriage in the London Transport Museum. *(London Transport Museum)*

A world first for London: the City & South London Railway

In 1890 London once again led the world with two ground-breaking civil engineering achievements: the first underground railway to use electric traction, and the first that would run not in a covered-over trench just below street level, but, almost uniquely, in tubular tunnels bored deep under London's clay, well clear of sewers, gas and electricity mains. Such a system of deep lines constructed in this manner cannot be found anywhere else in the world. In addition the City & South London Railway was the first underground railway with its own stations with lifts (necessary because of the depth of the platforms) and a proper form of signalling. The line, which initially ran between King William Street and Stockwell, would later form the nucleus of today's Northern Line City Branch.

In all, 52 small, stubby locomotives were ordered from a variety of manufacturers for the new line between 1889 and 1901. The first 14 were from Mather and Platt of Manchester, who also provided all of the electrical installations. They had recently formed an electrical engineering division headed up by Edward Hopkinson, whose elder brother, Dr John Hopkinson, had improved the Edison dynamo and was consultant electrical engineer on the project. The next two locomotives were from Siemens Brothers and three were built in their own works. They only measured 14ft in length over their centre buffers, weighed almost 10.5 tons and were mounted on four wheels. Each of the axles was driven by a 50hp motor, with its armature built integrally. The two motors were connected in series and took their power at 500dc through cast-iron collector shoes from a third rail; but this power supply was never really able to cope with several trains accelerating away at the same time.

One hundred and sixty-five wooden carriages were ordered between 1890 and 1907, the first 30 from the Ashbury Carriage and Iron Co, builders of the first carriages for the Metropolitan. The trains were short, with the locomotives hauling only three carriages seating just 32 passengers. The first series of carriages were extremely claustrophobic since they had tiny, narrow, slatted windows, it having been thought that passengers would neither need nor wish to look out of them. Not surprisingly they were dubbed 'padded cells'! However, further builds would have half-height and then full-height windows fitted. These last were also of all-steel construction, anticipating the Board of Trade's later stipulations regarding tube train construction.

The carriages were at first fitted with lattice gates at both ends for entry and exit. This unfriendly and intimidating solution was imported from America, and continued to be specified on subsequent tube train designs until a better solution was eventually found, not finally disappearing from service until 1927. 'Gatemen' rode on the carriage platforms and had to call out the names of stations to passengers, and for a number of years their verbal messages were augmented by enamelled station-name indicator plates that showed through a rectangular opening in the end doors of the carriages. One can only wonder how visible these would have been in a crowded train!

Twelve hydraulically operated lifts were originally provided at each of the six stations (Stockwell, The Oval, Kennington, Elephant and Castle, Borough and King William Street), with Kennington receiving an electric lift in 1897. The lifting gear was housed in a domed roof that at the time characterised the appearance of all the stations, but only that at Kennington survives

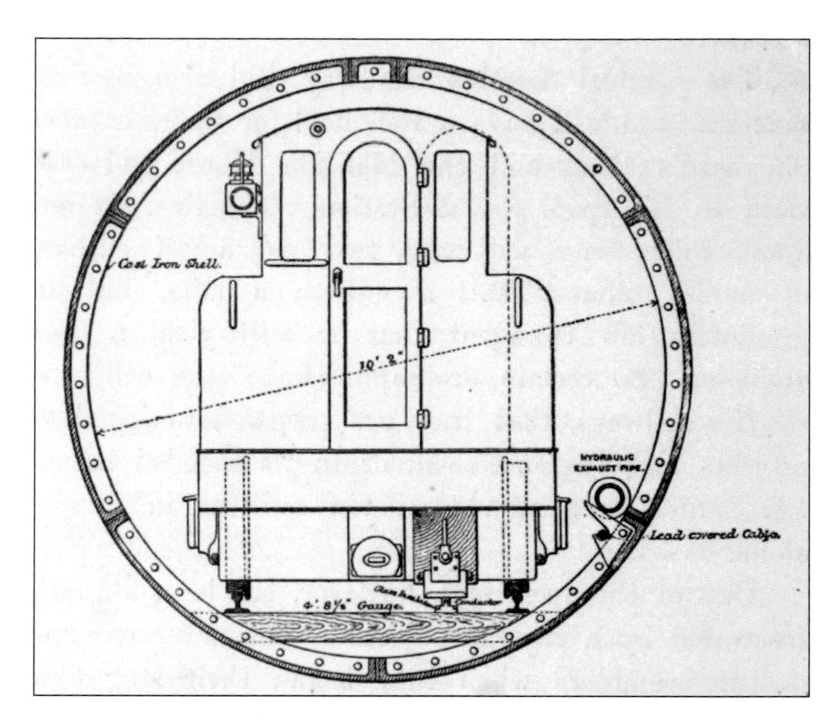

ABOVE Tight fit! Cross-section of the loco in its tunnel. *(Grace's Guide)*

BELOW C&SLR train in later years with deeper carriage windows. *(Mike Ashworth collection)*

ABOVE Kennington station as first built. *(London Transport Museum)*

today – the only one of the original buildings not subsequently replaced or substantially altered.

Although the Greathead shield, as the next stage of development of the Barlow shield was named, allowed tunnelling to be carried out far below all of the sub-level services and foundations along the route, the promoters were nevertheless still wary of having to pay out compensation, so the line's route still essentially followed the line of the streets above. In some cases this forced the tunnels to be 'piggybacked' on top of each other, giving an undulating riding experience not unlike a fairground roller-coaster, which threw standing passengers all around the car. Although the trains ran on the standard 4ft 8½in gauge, the very narrow tunnels and small-section carriages certainly did the fare-paying passengers no favours. The rough and bumpy ride, with no straps or handrails to grab, was made even worse by the fact that the electric lighting dimmed to the merest glimmer during power surges, and the overall smell of the line (as recalled by John Betjeman from his childhood days) had a foetid smell akin to a changing room – a commentary on the hygiene and personal cleanliness of many Edwardians.

The world's first underground electric railway was officially opened by Edward, Prince of Wales, on 4 November 1890, and opened to the public on 18 December. Unlike other railways, the City & South London Railway had no ticket classes or paper tickets; a flat fare of 2d was collected at a turnstile. In its first year of operation, despite the cramped carriages, the line attracted 5.1 million passengers, little more than half the number carried by the Metropolitan in its first year of operation, and the number grew very slowly, to 7 million in 1899. It was to be seven years before the company paid a meagre 2% dividend to shareholders, which certainly discouraged investment in further such schemes.

Unfortunately the line proved to have been built too small (its small tunnels were found to be quite inadequate to meet the demand), it was underpowered, poorly routed and never really profitable. Although full to the brim during the hours of commuter peak travel, it ran almost empty at other times, with the obvious outcome that it was not a commercial success.

However, despite all of its financial problems it still managed to raise enough capital to bypass its difficult curving tunnel into the King William Street terminus by building a new link from just north of Borough Station to Moorgate. By 1900 this reached northwards to the Angel, Islington, and south from Stockwell to Clapham. Further extensions in 1907 connected with Euston and King's Cross. One can only guess at how the line managed to cope with the immense congestion brought about by the continued use of its small locomotives and carriages during the First World War, and it must have clearly demonstrated that drastic and very costly improvements were required.

The C&SLR paid the price for being a pioneer in its field, its mistakes being identified and acted upon by subsequent underground train engineers and designers. Nonetheless, it had demonstrated that such railways had a future.

The Waterloo & City Railway

Of the five new tube railways authorised by Parliament between 1891 and 1893 only one was completed by the turn of the century, and this railway, built under the Thames, was significant in several ways. A Joint Parliamentary Committee had recommended that in future a minimum tunnel diameter of 11ft 6in (the revised size used on the C&SLR's extensions) should be specified and, more importantly, that underground railway companies should no longer be obliged to purchase any property under which they would pass. The legal principle of a 'wayleave' (permission to cross through someone's land) was now considered to be sufficient, and it suggested that this should be granted free of charge to any future lines running below the public streets.

The directors of the London & South Western Railway had long eyed the lucrative possibility of linking their main-line terminus at Waterloo, just south of the river, with the City, so that their well-heeled passengers from the leafy Surrey stockbroker belt could be swiftly delivered non-stop to their places of work. Clearly a new elevated surface route was totally out of the question, so a deep tube railway was the chosen solution. With this potentially lucrative market in mind, the Company guaranteed the required finance of £540,000 for construction of the short, mile-and-a-half line, ensuring returns of 3% to shareholders. It was built in just four years and opened in 1898.

A larger tunnel size of 12ft 1½in was chosen and the wooden rolling stock was ordered from Jackson & Sharp of Delaware in the USA, since it was stated that no British manufacturer could meet the very tight delivery dates. These were assembled at the Fastleigh carriage works of the London and South Western, and the electrical equipment was provided by Siemens Bros & Co Ltd. An order for five further motored cars was, however, placed with the British Dick, Kerr & Company of Preston, famous for its tramcar production. This time the trains were not locomotive-hauled but instead the sets were operated by powered driving motor cars at each end, with two electric motors driving each set of two wheels inside their four-wheeled bogies. The operating current of a nominal 600V came from a third central rail via a collector shoe and was distributed along the train's length to the end bogie by a roof-mounted power cable. The Board of Trade were naturally very nervous about such an arrangement for safety reasons, and forbade its use on any further train type. It was, however, a technical stepping-stone to the development of full multiple-unit traction, and the trains were considerably lighter and cheaper to operate.

ABOVE One of the first Waterloo & City line trains. *(London Transport Museum)*

The Central London Railway: the first modern tube line

The Central London was the true pioneer line of the Underground tube network; a step change was introduced in the train's operating technology, which has been carried through all succeeding types of tube rolling stock, and the line's deep level platforms were essentially what we still have today. From the outset the company promised the best financial returns for investors since it would run underneath London's busiest omnibus routes via Cheapside, Holborn, Oxford Circus and Bayswater Road to Shepherd's Bush.

This third London tube was a substantially larger enterprise than the preceding ventures, and would cost nearly five times as much to build as the original City and South London had. When it opened for business in 1900, the revenue taken from the Metropolitan's northern end of the Circle line and the District's southern end was clearly going to be considerably threatened, and this would accelerate the conversion of their steam-hauled rolling stock to electric traction in order to successfully compete with the upstart.

The financing of the line heralded the first direct involvement of Americans; they would also play a huge role in the technical development and design features of

Underground trains right up to the First World War, and their legacy lives on today. Present-day users of the Underground would be very surprised to know that the London tube system owes its very existence to American finance!

The Exploration Company Ltd was founded in 1889 to develop overseas mining interests and was backed by the Rothschild family. This company formed a consortium to promote the Central London Railway, and the syndicate owned just over half of the latter's shares. Other shares were held by extremely wealthy friends of the Prince of Wales, such as Sir Ernest Cassel, Henry Oppenheim and Darius Ogden Mills, all of whom had considerable stakes in American businesses and corporations.

Construction started in 1896 with the redoubtable trio of Sir John Fowler, James Henry Greathead and Sir Benjamin Baker supervising the engineering. Only Baker would live to see the line completed. In order to keep costs down while establishing the railway's route, the previously mentioned niceties of the 'wayleave' process were applied wherever possible, and street plans were followed as closely as was practicable.

Even travelling on the Central line today there are reminders of this, because in order to deal with a localised narrowing of the street route, lines would once again 'piggyback' each other rather than running parallel at the same depth. This is why three of today's stations – St Paul's, Chancery Lane and Notting Hill Gate – have their eastbound and westbound platforms at different levels.

As expected, tunnelling followed the same established procedures as before, and 48 electric rather than hydraulic lifts were installed at the 13 stations along the line. The ceiling vaulting and walls were fully covered with glossy white tiles, which together with overhead electric arc lamps created a very bright yet somewhat stark ambience that was nevertheless a great improvement compared to the rudimentary gas-lit City and South London platforms. Two of these early platforms survived in original, albeit by now fairly decrepit, condition at Lancaster Gate Station (previously known as Westbourne Station) until the early 21st century and were the last to be updated with fresh, new white tiles. Indeed, several other tube stations along the route have been similarly fitted out; a nod to the effectiveness of the original hundred-year-old illumination scheme. The wooden planks lining the platforms from the outset were replaced by stone paving slabs after a few years, following a major fire on the Paris Métro in 1903.

RIGHT An original white-tiled Central London Railway platform. *(London Transport Museum)*

LEFT Holland Park station today. *(Mark Ovenden)*

The external station elevations were a very early example of a 'corporate' solution; the external elevations were clad with unglazed terracotta in a warm cinnamon colour, and were all designed by the architect Harry B. Measures. They each involved decorative variations on a standard solution to suit the individual footprint requirements of any one site, and some still survive today. Holland Park is arguably now the closest to its original appearance.

The line was opened on 27 June 1900 by the Prince of Wales, and amongst the 'great and the good' present for the occasion were several notable Americans, including Mark Twain, who was living in London at the time. It was not until the end of July that it was opened to the public. As with previous openings, the line was an instant success, but this time it would remain so over the years. A flat fare of 2d had a lot to do with its success, and the line soon earned the sobriquet 'The Twopenny Tube', and was branded as such on its official posters. The great advantage that the line possessed over its predecessors was that it was not wholly reliant on commuter traffic, since it readily accessed London's main shopping venues as well as 'Theatreland'. This enabled it to attract a ridership of nearly 15 million by the end of 1900 and just over 41 million throughout 1901. This was eight times as much as the City & South London achieved in its first year.

As a result of these figures the Central was able to pay good dividends to its shareholders right from the start. A generous 4% was paid for the first five years of operation and then 3% from 1905 to the first day of 1913, when it merged with the Underground Group. There was, however, condemnation of the 'obnoxious and rank odour' experienced by passengers on the line and many experiments were made to satisfactorily ventilate the railway. It was not until 1911 that a complex pressure system was installed which pumped down ozonised air. Again John Betjeman recalled the Central London from his childhood days by recollecting that the Company tried to make its stations smell like 'Margate by the seaside', but doubted the efficacy.

Two types of rolling stock employed to haul trains on the line were American in both design and technology. Initially locomotive-hauled trains were used, and then of necessity replaced by multiple units. Although the British Thomson-Houston (BTH) company had secured the entire electrical contract for the line, the original locomotives were designed and built in America by the General Electric Company of Schenectady, New York (of which the aforementioned Darius Ogden Mills was a director), with whom BTH were associated.

Greathead initially wanted the trains to have locomotives at each end, connected by a continuous power cable as on Waterloo & City stock, in order to spread the tractive effort, but the Board of Trade would this time have none of it. So he had to settle for one very powerful

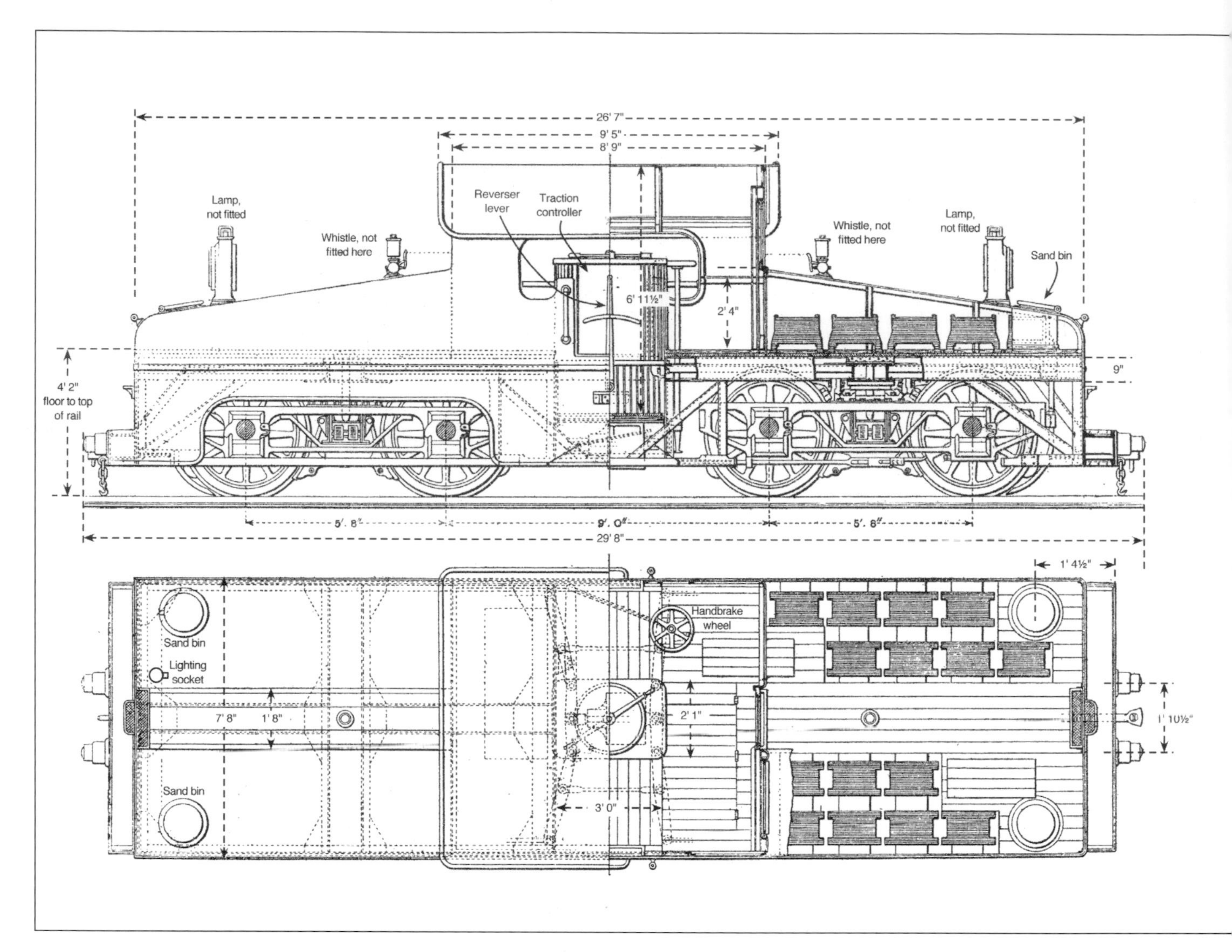

ABOVE A CLR electric locomotive. *(Piers Connor)*

unit instead, which weighed 44 tons of which around 33 tons was unsprung, as the motor armatures were once again built on the axles that in turn carried the heavy motor frames. Gearing for locomotive drives was in its infancy in 1900 and was judged to be too noisy for use in deep tube tunnels. As a consequence the huge sprung weight almost immediately caused serious complaints of excessive vibration from property owners along the route.

Enter Frank Sprague and multiple unit technology

What was the solution? By August 1901 experiments had been made to reduce the overall weight of the locomotives to 31 tons, just over 10 tons of the saving being achieved by fitting new geared nose suspended motors. These were a step in the right direction, but a better long-term solution had still to be found. The stage was now set for another American to come to the rescue, this time the extremely talented engineer and inventor Mr Frank J. Sprague. By 1889 110 electric railways had or were about to incorporate Sprague's electric equipment on several continents. Edison, who manufactured most of this apparatus, bought him out in 1890, after which he turned his attention to electric lifts, a far superior operating system than hydraulic power. In 1892 he had formed the Sprague Electric Elevator Company and had developed systems to separately and safely control a number of lifts, in particular for the then ever-increasing number of skyscrapers being built in America. After the worldwide installation of 584 of his lifts worldwide, he sold his business to the Otis Elevator Company in 1895 for a handsome

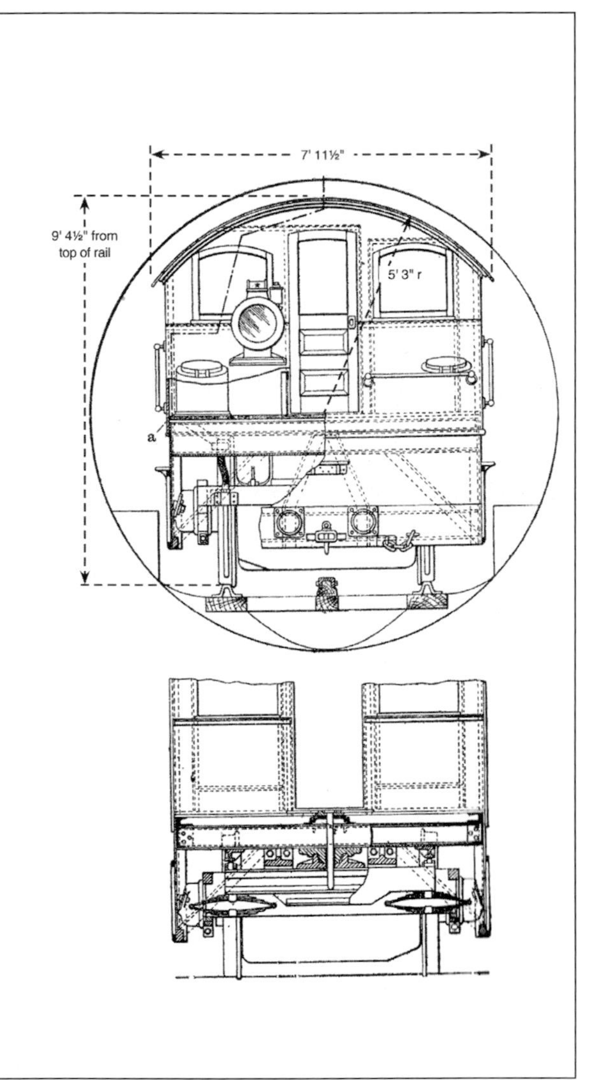

profit, and they subsequently received the order for all of the Central line's lifts.

In an astute example of lateral thinking (literally!), Sprague turned the concept of his lift control equipment through 90° to invent a multiple control system of railway operation that greatly accelerated the development of electric traction. In this system each carriage (henceforth called a 'car' in this book, from the inherited American parlance) of the train carried its own electric traction motors. By means of relays energised by the available trackside current, the driver could command all of the motors to act together, the control being achieved by employing a number of low-current electrical circuits that remotely operated the control switches that made and broke the main power circuits. Thus power could be distributed right along the train's length, and at a stroke the previous solution of a continuous power cable operating at full voltage was eliminated.

As a result four of the existing trailer cars were converted to multiple-unit operation, with one bogie at each car-end replaced by a power bogie with two 100hp motors. These cars entered service in September 1901 and were an immediate success. The vibration was virtually eliminated, and the multiple units had the added advantage of much greater flexibility at termini, since with a cab fitted at both ends trains could now be easily driven in either direction, whereas locomotive-hauling had previously entailed turning either the loco or the whole train around via a loop line. Finance was soon found to order 64 new motor cars from two British manufacturers, and the complete change over to multiple-unit traction had taken place by June 1903.

TOP Rear of a trailer car converted to multiple-unit operation. *(Bob Greenaway collection)*

ABOVE The 'business end'. *(Bob Greenaway collection)*

ABOVE Interior view of car. *(London Transport Museum)*

Car interior design rapidly went through several permutations; the original perforated wooden seats were soon replaced by sprung rattan weave (split cane), but the very elegant art nouveau-inspired clam-shell light fittings with an elegant intertwined CLR motif were also soon replaced by more perfunctory glass shades, probably to increase their light output. The cars were very smartly finished in brown (with a purplish tint) and white with 'Central London' emblazoned on the sides in 4in-high gold letters with the company's coat of arms. Entry to the vehicles was again via end gates because a better solution had yet to be developed.

Operating the trains

The operation of these trains was extremely labour intensive and would remain so until a major rebuild was carried out in the mid-1920s to convert them to air-powered doors. After this major rebuild of the passenger cars in 1926 they then continued in service until June 1939, an amazing feat considering that they were still using the same operating technology that had been rapidly incorporated 36 years previously.

Originally a crew of eight was carried on a seven-car train: the driver with his assistant, a front guard and a rear guard and a further four gatemen in between. The starting procedure went like this (it must have been extremely difficult to manage, with the crowded platforms of Edwardian times). Each gateman had to face forward when his gates were shut. When the front guard saw the correct number of hands (one for each gateman), he showed a green light, to which the rear guard replied by showing his green light forward and blowing a whistle. After he had received these various signals, the front guard showed his green light to the driver

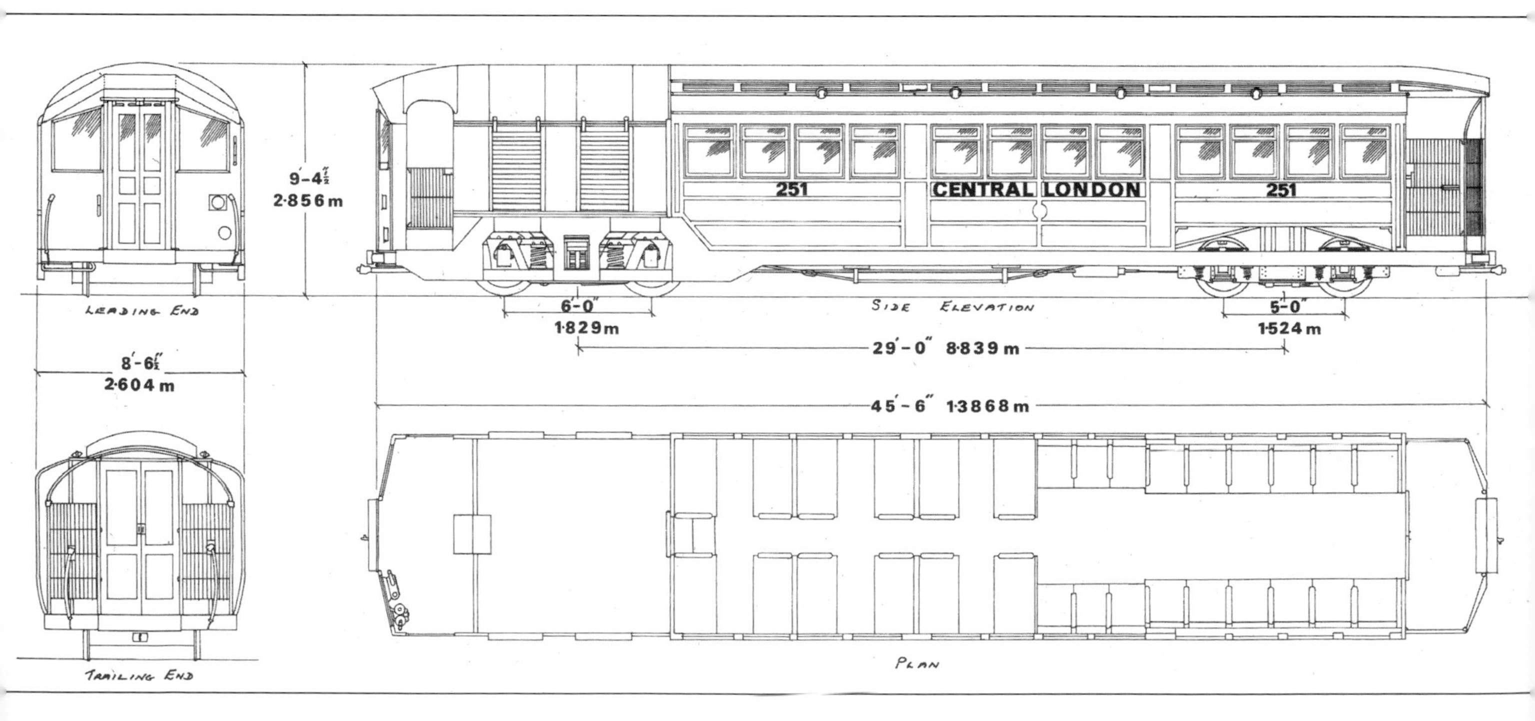

BELOW CLR multiple unit elevations. *(Piers Connor/LURS)*

(or his assistant) and finally the train could start. What a way to run a railway!

After 1905 The Central London suffered a steady drop in ridership as a result of Circle line electrification and the proliferation of reliable petrol- and steam-driven motor buses along its routes. From July 1907 the 2d flat-fare policy was discarded and replaced by 3d for passengers travelling more than eight stations from their departure point. Short extensions at either end were later added. In 1908 a half-mile loop linked Shepherd's Bush to Wood Lane, which provided an entry point to 'The White City' exhibition grounds – the 140-acre site of the Franco-British Exhibition with its 100 elaborate Indian Palace-style buildings, fashioned from white-painted fibrous plaster with surrounding boating lakes. An adjacent stadium successfully hosted the Olympic Games in the same year after Italy pulled out at the 11th hour due an eruption of Mount Vesuvius.

The fact that the exhibition itself attracted eight million visitors in just six months, plus the spectators for the Olympics, brought a brief respite to the Central's fortunes, though their decline resumed the following year. Three other major exhibitions in 1910 (the Japan British), in 1912 (the Latin British) and in 1914 (the Anglo-American) would again temporarily stimulate ridership, but a far more important link would be finally made in 1912 to extend eastwards from the Bank and connect with the Liverpool Street and Broad Street main-line stations.

LEFT CLR poster for the 1912 Japanese Exhibition. *(London Transport Museum)*

'The problem child': the Great Northern and City Railway

This was the first tube line to be built in the actual Edwardian era and was unique in that it was built with large 16ft diameter tunnels that could accommodate trains of full-size standard suburban stock. No other such master plan was to be specified in Britain for over another 100 years, until the huge 21st-century Crossrail undertaking that will once again see full-size trains running in large circular tunnels beneath London. The similar Parisian RER (*Réseau Express Régional*), a system that's similar in scope, had already pointed the way ahead during the 1970s. Although this French railway eventually grew into five major lines radiating in all directions from the city centre, its GN and the CR lines, like the original East London line, never had a significant routing with appropriate interchanges, and as a result eventually withered and died on the vine.

The Great Northern and City line was originally sponsored by the Great Northern Railway (GNR), which was the main line through to London on the east coast route to King's Cross. As with other main-line operators it wanted direct access to the City but was not able to achieve it by a surface railway, which would, as before, have required extensive property purchase and demolition. They were also experiencing difficulties in running their busy suburban services from Finsbury Park to King's Cross and over the previously mentioned 'City Widened Lines' to Moorgate.

A tube line was therefore proposed to relieve the intense rail traffic congestion caused by this bottleneck. The plan was that steam-hauled suburban trains would leave the main line at Finsbury Park and run underground to

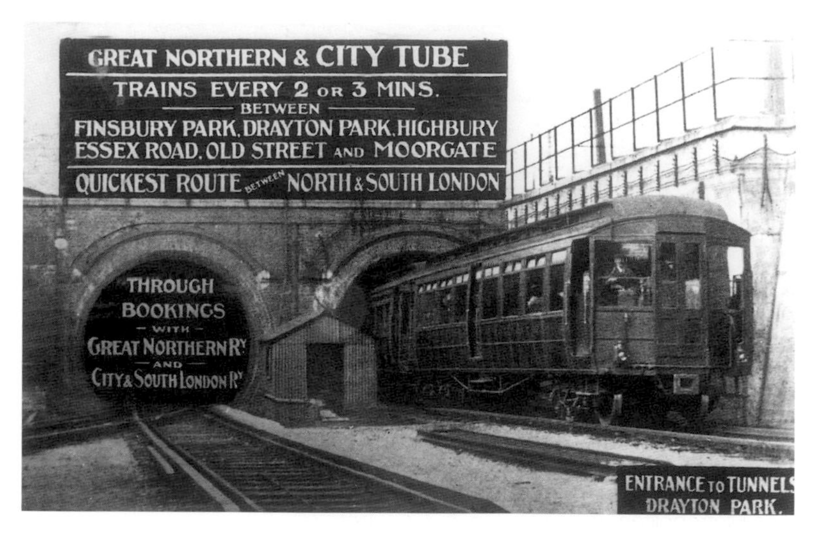

ABOVE Great Northern & City train exiting the tunnel at Drayton Park. *(LURS collection)*

Moorgate, having changed to electric traction at Drayton Park – hence the reason that the large-diameter tube tunnels were chosen, so that they could accommodate standard suburban-sized carriages.

Construction of this 3½-mile line was authorised by Parliament in 1892, but the GNR, in spite of its initial backing, refused to provide any financial support for the project. Funds were eventually garnered from other sources, including a significant investment from S. Pearsons & Sons, the contractors for the line, who bought many of the shares and guaranteed a 3% dividend to get the line built. The main tube section between Drayton Park and Moorgate was completed in 1902, but by then the Great Northern had lost all interest in its original scheme, to the point where running rights into its Finsbury Park station were withdrawn. Consequently the embryonic line had to make do with its own separate small underground station that had an inconvenient interchange. As a result the original hopes to relieve pressure on the Great Northern's suburban network came to nothing, since through working by their trains was impossible because the proposed junction with their main line near Drayton Park was abandoned. Pearsons themselves ran the line for three loss-making years and then sold it to the Metropolitan in 1903.

BELOW A train at Finsbury Park station. *(LURS collection)*

The Metropolitan likewise never developed the line's initial promise but half-heartedly kept it going in splendid isolation for a further 20 years, when it was absorbed by London Transport following nationalisation. The original rolling stock was withdrawn in 1939 after which Tube Stock took over and it subsequently became part of the Northern line. In 1975, shortly after the Underground's horrendous worst-ever peacetime disaster at Moorgate, when a train of 1938 Tube Stock ploughed into a dead-end tunnel with the loss of 41 lives, the line was transferred to British Rail. It was then converted back to take full-size trains once more with, finally, a proper connection to the suburban network at Finsbury Park. Thus the line's original objective was finally realised some 70 years later.

The styling of the rolling stock was very American in appearance and, with their clerestory roofs smoothed towards the cab fronts, was very similar to the new electric trains on order for the District line. However, rather than a bluff front end they featured an attractively facetted treatment. A total of 32 motored cars and 44 trailer cars with gated ends were also ordered from Brush and the Dick Kerr Company. They were fitted with the Sprague/Thomson-Houston multiple unit control system, ordered before this arrangement was also specified for the Central London Railway; however, the CLR vehicles were the first to be put into service. Fifty-eight teak-bodied cars were available for the line's opening in 1904, with the remaining 18, of all-steel construction, delivered two years later.

Of historical importance is the fact that this was the first tube railway to not only have automatic signalling throughout, using light signals with no moving parts, but also to have an insulated traction return current.

Chapter 4

Electrification of the sub-surface lines

By the late 1890s electric traction technology had developed to the point where trains with an all-up weight of 100 tons could be successfully operated. As well as the technical success of the three tube lines, developments in both Europe and America were also providing examples of the way forward. For example, Budapest had opened its first cut-and-cover metro line in 1896, the first in mainland Europe. This utilised the Ganz system of three-phase alternating current drawn from an overhead supply. The rolling stock employed was not unlike that used on the Waterloo and City line, with a power car carrying two motors on top of each axle. The first 20 cars were built by Siemens and Halske, and remarkably would continue in service until 1973, when they were almost 80 years old! Paris and Berlin were also developing their own fully electrified metro systems that would open in 1900 and 1902 respectively.

Meanwhile in London, electrifying the Circle line would receive the greatest priority, as it had lost a lot of custom to the CLR. The era of steam haulage was finally dead, and wholesale scrapping of the locos would be carried out as soon as practicable. But which was the best operating system to specify?

In 1898 the Metropolitan and the District railways had agreed to share the cost of an 11-month trial with Siemens electric traction on a four-rail system between Earls Court and High Street Kensington, just one stop away. The experimental train was composed of six cars, the cost of which was split between both companies. This was obviously a success, because both parties then set up a joint committee that fully recommended electrification but did not endorse the system that was trialled. Tenders were therefore submitted, and the Budapest Ganz 3,000V three-phase ac system using a similar overhead wire supply was the recommendation; cost must have been the overriding factor, since it was cheaper than a conductor rail solution. The Metropolitan was in favour, but just before the go-ahead was rubber-stamped another American, this time a particularly powerful one, arrived on the scene – one Charles Tyson Yerkes, of whom more anon.

Yerkes had recently effectively secured ownership of the District via a management buy-out, and had turned down the Ganz system in preference for a low-voltage dc arrangement. This was the preferred choice in America, and he already had it in mind as the choice for the three tube railways that were at this stage still a gleam in his eye. The Metropolitan challenged the ruling, which then went to arbitration. However, they lost their appeal when the Board of Trade ruled in favour of the Yerkes-sponsored four-rail dc arrangement. Since a

LEFT Charles Yerkes. *(London Transport Museum)*

ABOVE A 1905 Metropolitan electric train. *(Bob Greenaway collection)*

fully integrated traction system was essential for operating the Circle line, the Metropolitan had no option but to accept the findings and then shut up. They firmly rejected a take-over by Yerkes at this juncture and proceeded to electrify using a 600V system as specified by the District (third and fourth rails, with the positive rail mounted outside the running rails with the negative one in the centre). They built their own generating station at Neasden, which opened for business in December 1904 and one month later started to run their first multiple-unit service between Baker Street and Uxbridge via Harrow. A newly built branch from Harrow to Uxbridge had opened in July 1904 but had to be steam-hauled until electrification was completed.

Charles Yerkes and other Americans

The District's electrification was carried out by their chief engineer James Russell Chapman – another American in the saga – and the whole line was converted by the end of 1905. It had cost £1.7 million to complete (£135 million in today's money). Power was drawn from a huge generating station built at Lots Road, Chelsea, near the Thames, between 1903 and 1905. A constant supply of coal from barges was very conveniently managed and at the time it was the largest electrical power station in Europe. Henry Ford had already established the concept of river-fed coal supplies to his manufacturing plants, first on the River Rouge at Detroit, then on the Thames at Dagenham, and finally on the Rhine at Cologne. The sheer size of his power station fitted well with Charles Yerkes' aspirations to additionally supply three nascent tube lines as well as a tramway system on which he had set his sights.

Not only was he the driving force behind the District's electrification, he was also instrumental in the construction of the core of today's underground network, that is to say the Bakerloo, Northern and Piccadilly lines, which still serve hundreds of thousands of today's passengers daily. It was Yerkes alone who ensured that London had its own tube network,

RIGHT Lots Road generating station. *(Capital Transport collection)*

and without him it would probably never have been built at all. Although plans to build these lines had already received Parliamentary approval, drawn-out planning and financial difficulties meant that they had never got beyond first base until Yerkes appeared on the scene.

As a young man in his early 20s he had already made as fortune as a bond dealer. However, he lost the lot following the Great Chicago Fire of 1871, when insurance companies all over America defaulted on the damage to the large commercial district. He was later jailed for several months for alleged embezzlement of funds. After that he got involved in developing Chicago's tramway and suburban railway systems through shrewd and complex business dealings. These became increasingly more tarnished and dishonourable until he was ultimately forced to leave Chicago, having failed to renew franchises on his transport undertakings.

Pocketing his personal fortune of $15 million in cash, he moved to New York, where he indulged in fine living and dining in a Park Avenue mansion with a string of mistresses. It was while he was residing in New York that he was contacted by R.W. Perks, the solicitor of the Metropolitan Railway, who persuaded him to finance the electrification of London's District Railway. By 1898 Yerkes had already bought a large interest in the District Railway, recognising its dilapidated condition as an opportunity for renewal, which electrification would certainly offer. He was also particularly interested in three proposed deep-level tube lines that were floundering for lack of finance.

The first of these was the Baker Street and Waterloo Railway (soon abbreviated to the sobriquet 'Bakerloo'). It was planned to be three miles long, with intermediate stations at Oxford and Piccadilly circuses that would later become major hubs on the network. Ironically, the initial background to the financing of this line demonstrated that Americans did not have a monopoly where fraud and embezzlement were concerned in the building of municipal railways.

Authorised in 1893, the Baker Street and Waterloo Railway was struggling to raise the necessary capital when Whitaker Wright, an English millionaire who had made a fortune from mining interests in America, approached its management with an offer to invest £700,000 in the line, funded by the London & Globe Finance Co, in which he owned almost one third of the shares. Construction started in 1898 but stalled 18 months later when Wright's finance company failed. Work stopped as contractors queued up to be paid. Wright then fled the country bankrupt, but was hauled back to London from New York to face the music. In 1904 he was sentenced to seven years' penal servitude for defrauding investors to the tune of £5 million. Just minutes after protesting his innocence he took a cyanide capsule, and expired at the Royal Courts of Justice. Quick off the mark, Yerkes' consortium bought the Bakerloo line early in 1902.

The next emerging line to join his portfolio was the Charing Cross, Euston and Hampstead Railway (now part of the Northern line). He obtained Parliamentary authorisation to build the line and in September 1900 invested £100,000 to kick-start its construction. He became its chairman and appointed Perks as director. Yerkes had in his Chicago days exploited the real estate potential alongside his suburban lines, and of course this same philosophy was applied to London, particularly in the more affluent parts. Thus Golders Green was planned as the terminus, with the tube line running under the Heath. The station at Hampstead would become the deepest on the entire system, its platforms being 192ft (58.5m) below sea level, with its deepest lift shaft measuring 181ft (55.2m).

Having secured control of the District in March 1901, Yerkes formed an umbrella holding company four months later, called

ABOVE The power generating equipment. *(Capital Transport collection)*

the Metropolitan (a strange name choice, because he had no control over that company whatsoever) District Electric Traction Company (MDET), which only nine months later was reformed as the Underground Electric Railway Company of London (UERL), thus ditching the anomaly of the word 'Metropolitan' in its title. The third tube line to be built by the Yerkes Group was the central section of today's Piccadilly line, created by the combination of two separate railways that were also on the stocks. These were the Brompton & Piccadilly Circus Railway and the Great Northern and Strand. They were both acquired in September 1901.

ABOVE Chalk Farm. *(Author's collection)*

Yerkes' stations and trains

Within a scant period of just five years all three of these lines were in operation, served by a grand total of around 50 buildings at 43 sites. The stations numbered 38 in all, seven with double frontages while Piccadilly Circus had three. The others were all entirely underground. These were all designed by the group's architect, Leslie W. Green, who brilliantly used a modular 'kit of parts' approach to the demands of each site. He tragically died of tuberculosis at the age of 33, no doubt exacerbated by his tremendous workload. His stations varied in layout from the large V-shaped corner building at Chalk Farm to more typical stations such as Belsize Park and Covent Garden.

Many survive in excellent condition to this day, and they're still a distinctive feature of London's streetscape. They were built using a load-bearing steel frame construction – a building system still very much in its infancy in Britain at the time, but, again, very common in American cities such as Chicago and New York. They were clad with glossy, ruby red moulded faience tiles, with distinctive gold lettering, and their interiors were characterised by ornate tiled ticket windows in the art nouveau style, which Green followed through to other details such as the grilles over the lift vent shafts.

Unfortunately many of these items were destroyed during station modernisation projects in the early 1980s and the construction of the Underground Ticketing System's secure suites

BELOW Early photograph of Belsize Park. *(London Transport Museum)*

BOTTOM The same building in recent years. *(Bob Greenaway collection)*

at the end of the same decade. A few precise replicas of the ticket windows were made for some stations, but it is only at Holloway Road station that the Edwardian originals can still be seen – fortunately the particular dimensions of the new secure suite there enabled them to still be used.

Unlike the platforms of the CLR with their overall glossy yet bland glazed white tiling, the Yerkes tube stations used white and cream tiles on the platform walls, highlighted by vibrant bands of contrasting colours; these were also employed on the stairways and passageways leading to the platforms themselves. These were also used to emphasise the form of the platform tunnels by arching over them at regular intervals. Each station could be readily identified by its own unique design and colour scheme. Station names in large, dark brown serifed letters were fused into the tiles located at both ends as well as at the middle of each platform. These names would some 30 years later be

LEFT Surviving original Leslie Green ticket window at Holloway Road station. *(Author's collection)*

LEFT Recently restored 1907 platform tiling. *(Mike Ashworth/ London Underground)*

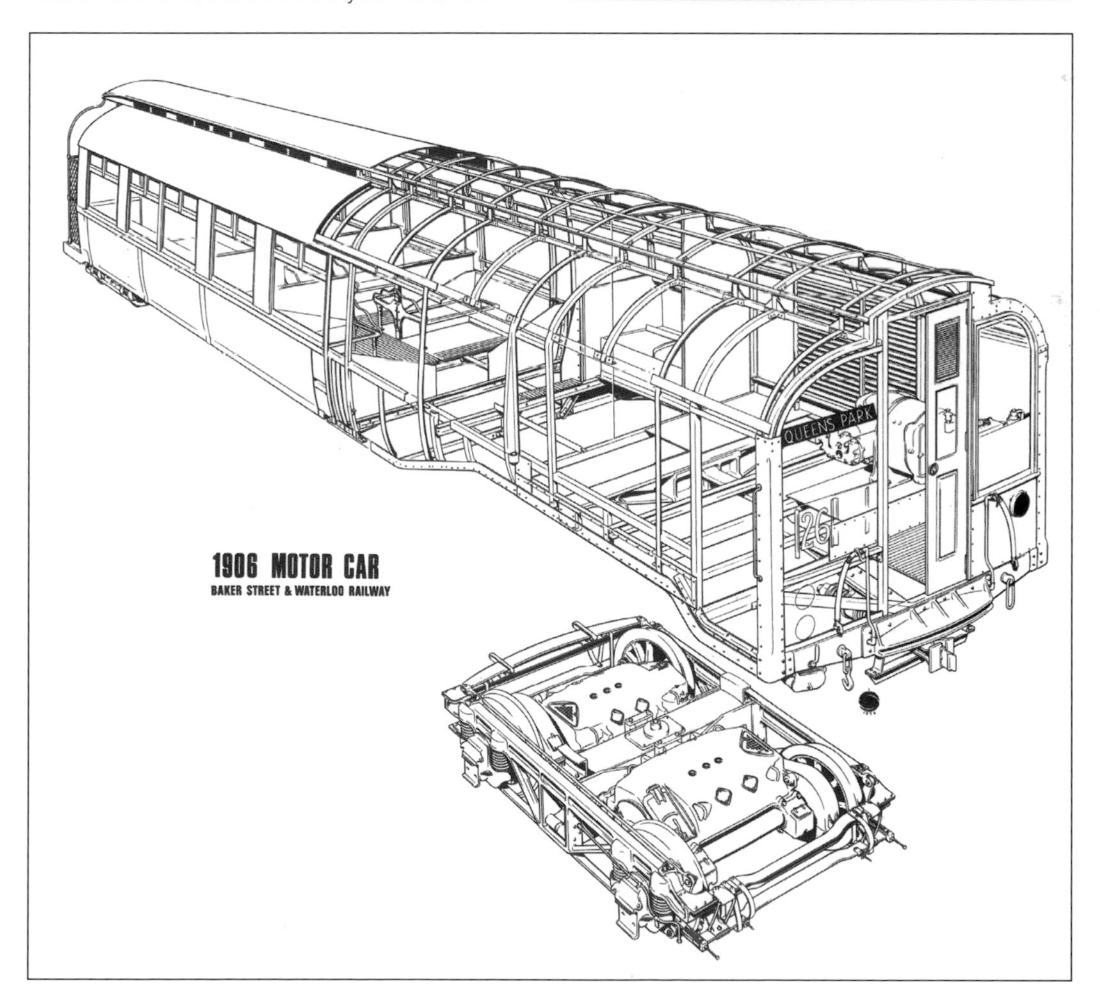

LEFT Cutaway drawing by Stewart Harris of a 1906 Yerkes driving motor car. *(Susan Harris/Andy Barr)*

ABOVE The interior of a typical Yerkes vehicle. *(Capital Transport collection)*

augmented by a continuous name frieze, originally using thick paper, later replaced by vitreous enamel.

Many of these original platform designs still remain today, and in recent years have been extensively (and expensively!) restored. A fine job has been done in matching the original coloured tiles and the quality of the tiling itself; indeed, it is now virtually impossible to detect the Edwardian originals from their restored counterparts. At the time of writing, the platforms of Hyde Park Corner, Holloway Road and Caledonian Road still retain their original platform tiling and it is easy to visualise them full of their original Edwardian passengers from 105 years ago.

BELOW The exterior. *(Capital Transport collection)*

Yerkes' tube trains once again used the CLR's Sprague multiple-unit technology with gated ends and no centrally located doors, one gateman being available to manually control each set of both car ends. Their job was made all the more difficult because, surprisingly, passengers were allowed to stand on the open gated platforms between stations to escape the cars' hot and stuffy atmosphere in the summer months. This would definitely be a non-starter in today's Health & Safety culture!

The hand signals used to give the 'all clear to start signal' on the CLR's trains were replaced by a system of ringing bells. The gateman located at the end of the train rang a bell through to his colleague once his group of passengers were safely on or off the train. This ringing of bells then proceeded along the length of the train until it reached the guard, who was positioned at the front between the first and second cars. Only when he sounded his own bell signal to the driver could the train move away. Amazingly, headways of around two minutes between trains were still possible with this laborious and archaic system. Technological developments were waiting in the wings to drastically improve this situation, as well as waiting times, but their introduction would have to wait until the end of the forthcoming First World War.

As a result of the previously mentioned horrendous fire on the Paris Metro in 1903, when 84 passengers died in a crush as they tried to escape the fire that raged through their train's softwood-bodied carriages, the UK's Board of Trade insisted that only steel-bodied cars with a minimal amount of internal hardwood features were to be procured for Yerkes' trains. Those used on the Bakerloo came from manufacturers in the USA; the rest of the fleet was obtained in equal quantities from France and Hungary, while only two trailers were built in Britain. They were all virtually identical in appearance, and the accompanying coloured Edwardian postcard images of a Piccadilly line vehicle provide an excellent impression of how they looked.

The excuse made for buying only two British cars was that British manufacturers did not have sufficient experience in building steel vehicles, but the order for the GN&CR's vehicles in 1906 shows that this argument was fallacious. No, the simple truth was that the prices quoted by British suppliers were up to 30–40% higher

LEFT 1903 'A' Stock train. *(London Transport Museum)*

than those of American and Continental manufacturers. (As a result British manufacturers would soon lower their prices, and of the new electric trains ordered for the District line 280 cars came from Europe but 140 were built in Britain, albeit to an American design.)

The Bakerloo vehicles were built by the American Car & Foundry, who bought premises in Trafford Park, Manchester, for their final finishing and installation of the electrical equipment. This was also to an American design, by General Electric, but was manufactured in England by their associated company British Thomson-Houston. They also built 108 cars for the Bakerloo and 150 cars for the Hampstead line. Half of the Piccadilly cars were built in France by Les Ateliers de Construction du Nord de la France (ANF) of Blanc Misseron; the other half were built by the Hungarian Railway Carriage & Machinery Works at Raab (now Győr). The interior design layout followed the pattern that had been established by the CLR cars, but they were rather more austere, with similar rattan-covered seats but perfunctory armrests.

A total of 140 lifts were installed on all three tube lines by the American Otis Elevator Company, and the American connections were further underlined by the fact that the signalling system and all the rest of the required electrical equipment were constructed to the plans already devised by the UERL's American chief engineer, J.R. Chapman, for the District line. One of his many innovations was to specify steam-powered turbines in the Lots Road Chelsea 'monster', to generate the required current of 500–600Vdc that would power the trains – one of the first power stations in the world to exclusively rely on such equipment. There were separate positive and negative conductor rails, the positive being outside the running rails and the negative between them. This layout would eventually become the standard arrangement throughout the London Underground system, but it would take right up to the beginning of the Second World War before such standardisation could at last be achieved. This was because the CLR, C&SLR and the GN&CR all had their own individual current collection systems.

A further all-embracing American association was the automatic signalling system used on all three tube lines. This employed track circuits to indicate when a train was in a particular section of a line, and a mechanical train-stop device that automatically stopped trains dead if they went through a red signal. These had been developed from the groundbreaking installation on the Boston Elevated Railway in 1901. Manufactured in England by the Westinghouse Brake & Signal Company, they had been first used on the District's Ealing and South Harrow line in 1903.

The District line's first electric trains

As the accompanying photograph shows, the first 'A' Stock trains of 1903 were, predictably, totally American in their styling, with

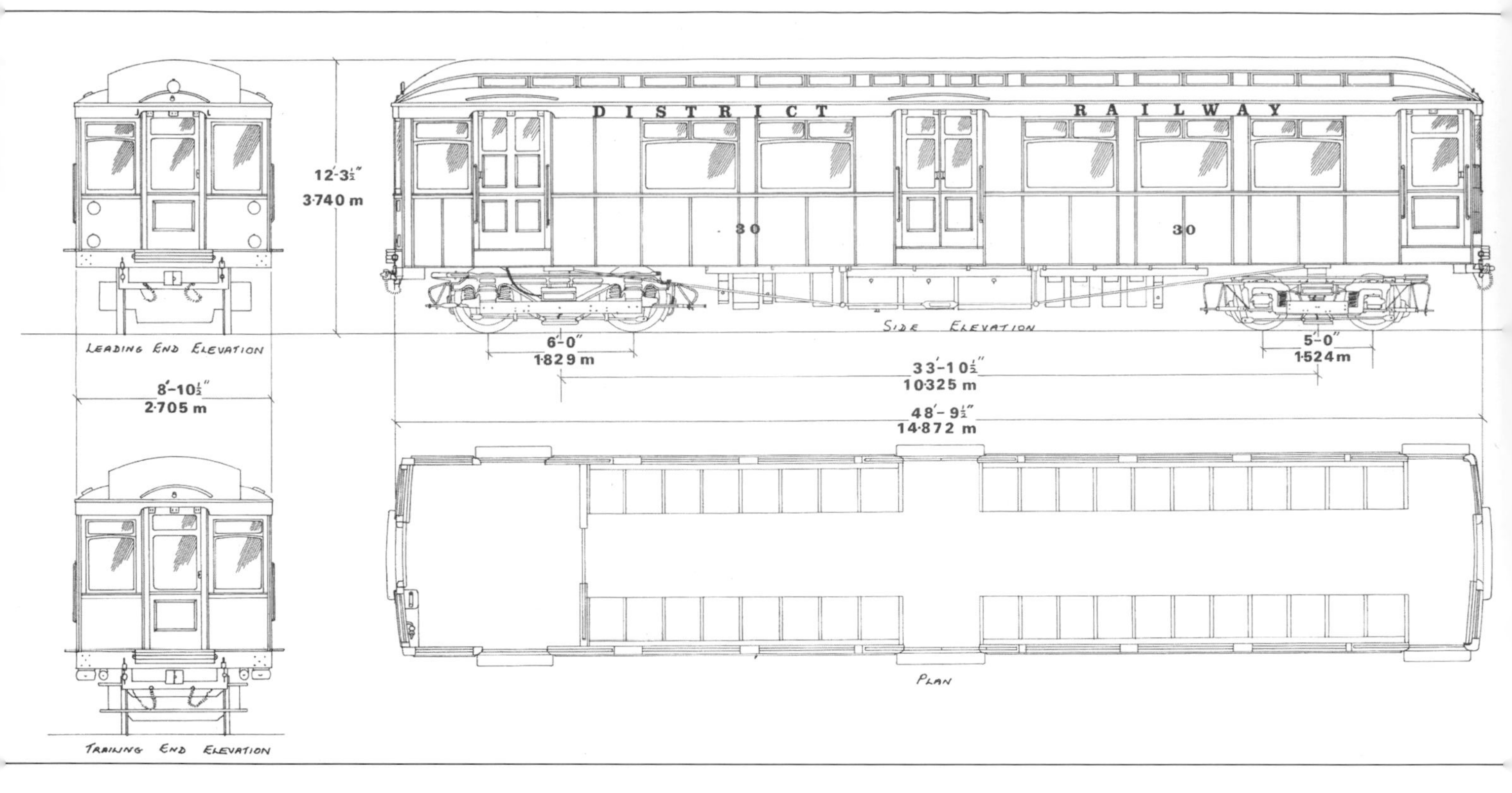

ABOVE 'B' Stock elevations. *(Piers Connor/LURS)*

BELOW Typical 'B' Stock interior. *(LURS collection)*

gated ends and arched windows. Although built by the Brush Electrical Engineering Co of Loughborough, they bore a striking resemblance to the vehicles built about the same time for the Interborough Rapid Transit Company of New York. They lasted in service until 1925, but the next type, the 'B' Stock – again ordered from Brush – had a far longer service life. Entering service in 1905, their appearance was as American as ever, with the name 'District Line' emblazoned down the sides of the cars in elongated American Railway serifed type. In a nutshell, they were an American transplant running on British tracks.

Further sets were ordered in 1911 and 1912 and the last of the type – still with its hand-worked door – was finally withdrawn in 1953, on the East London line. Not bad going for a train design whose technological design harked back to the very early days of electric traction. They were unique in that they had the distinction of serving Londoners throughout both world wars, and their longevity would be repeated by other Surface Stock types through succeeding decades, right down to the recently withdrawn Metropolitan line 'A' Stock, which was in service for just over half a century.

It's interesting to speculate what today's Health & Safety inspectors would have thought about the 'B' Stock's hand-worked doors, because passengers used to keep them open in hot summer weather for much-needed ventilation, and some people would even make a running jump to board a train through them when it was moving away!

Today's London owes a huge debt to Yerkes and his fellow investors, who were also mostly American; put bluntly, without them and Yerkes' retinue of American engineers London would possibly never had got either its tube system or its electrified District Railway. Money for these enterprises certainly wasn't forthcoming from the London Stock Exchange or a significant number of British financiers. Electric railways operating underground were still very much an unknown quantity to them, with many associated risks, notwithstanding the success of the CLR. A roll call of the most significant American mechanical and electrical engineers involved in the project reads as follows:

- James R. Chapman, who served for several years as general manager and chief engineer of the UERL.
- Zac Ellis Knapp, an electrical engineer from North Carolina who stayed to become naturalised and a director of the UERL.
- F.D. Ward, rolling stock engineer.
- S.B. Fortenbaugh, an electrical engineer loaned by the huge General Electric Co. of America.
- W.E. Mandelick, who became secretary of the UERL.
- G. Rosenbusch, assistant engineer, lifts and ventilation.
- J. Harrison, permanent way engineer.
- J. Thompson, electric lighting and telephones.
- J.W. Towle, power supply.

The London Underground Railway system was indeed incredibly fortunate to have such a vast fund of knowledge and experience at its disposal during its formative years. By contrast, Berlin and Paris never had such a head start, yet they still were able to develop their own truly excellent metros utilising their own home-grown talents.

Yerkes was remarkably successful in raising the necessary finance to build his lines. He had joined forces with the international banking firm of Speyers, which had offices in London, New York and Frankfurt. The London office was headed up by Sir Edgar Speyer, who promised to help Yerkes amass the sum of £5 million that was needed. The early 20th century was a period when a significant amount of American money was crossing the Atlantic to invest in British enterprises, which offered more opportunities for profitable investment than at home, and the investors whom Speyer and Yerkes approached certainly seemed as convinced as they were that fortunes were to be made by investing in London's tube railway network. Consequently 60% of the shares were sold to Americans, principally from established financial houses in New York and Boston, with the rest taken up by investors in Britain, France, Germany and Holland. The £5 million sum, however, proved nowhere near enough to fund the three tube lines and the District line's electrification, there being a shortfall of £10 million.

Yerkes managed to raise the difference by offering the existing shareholders 5% profit-sharing secured notes at 96%, redeemable at par on or after 1 June 1908. By this date he fully expected some profits would be available for distribution. This scheme proved to be very attractive, and about half of it ($16.5-million worth) was taken up in America. To top the amount up, 4% unsecured bonds were issued, safeguarded by mortgages on the Lots Road power station and the tube lines already partly built. All of these amounts added up to the required £15 million. However, all of these investors were to lose a lot of money, because, unlike the CLR, ridership and traffic receipts on all three lines were from the beginning well below Yerkes' most optimistic forecasts (about half of the predicted totals), and thus were never able to show the kind of return that he had promised his backers.

Although he lived to see his 'American' District line electric trains in action, Yerkes never saw his three tube lines in full operation. He died at the end of December 1905 at the age of 68, at the Waldorf-Astoria hotel in New York, as a result of kidney disease. His legacy, however, remains clear. Without his dynamism and financial wizardry (somewhat tainted though it was, to say the least) London's central tube network would never have progressed beyond thousands of rolled-up drawings on shelves collecting dust, and a scattering of half-completed, moribund station sites.

Stanley and Pick, the 'formidable pair'

To fill the void left by Yerkes' death, the directors of the UERL appointed Edgar Speyer (who from the beginning had guided the company's fortunes) as chairman, while a new face, Sir George Gibb, became deputy chairman and managing director. Gibb's previous experience and mature judgement marked him out as an excellent choice for the role, as he had been general manager of the North Eastern Railway, known for its highly professional management practices. Since a large proportion of the UERL's shares were held by American interests, they viewed their declining values with alarm and urged the appointing of a general manager with American drive and know-how to move the company forward. Gibb, with remarkably far-sighted vision and business acumen, appointed two men who in time would become known as 'the formidable pair'. They would work in concert to develop all aspects of the Underground network to become the world-beater against which all other metro systems would be measured.

In February 1907 Albert Stanley was appointed general manager at the then extraordinary salary of £2,000 a year, equivalent to well over £2 million today. Although born in Britain, his family had emigrated to America when he was a child, and he started to work on the Detroit tramway system at the age of 14, first as a messenger boy and then as a 'ganger', the foreman of a group of manual workers. His innate administrative and managerial strengths were soon noticed by his superiors, because at 18 he was made a divisional superintendent. Only two years later he became the overall controller, with responsibility for the opening of many new lines. Having moved to New Jersey in 1904, in 1907 he was made accountable for all of the state's transport undertakings, and soon afterwards he was offered the top job in London. It would appear that he initially regretted the move so soon after his promotion in America, particularly when he discovered that the UERL was haemorrhaging cash in a big way.

Ticket prices were too low and passenger numbers were significantly below the original pre-opening estimates. This was partly due to competition from the Metropolitan and Central London railways, as well as the increasing reliability of the capital's numerous petrol- and steam-powered motor bus fleets – which, ironically, were also very profitable.

Gibb's other able lieutenant was one Frank Pick, who had been his personal assistant at the North Eastern Railway since 1904. Born in Spalding, Lincolnshire, the son of a draper, he had qualified as a solicitor in 1902 and

RIGHT Albert Stanley, in later life Lord Ashfield. *(London Transport Museum)*

FAR RIGHT Frank Pick. *(London Transport Museum)*

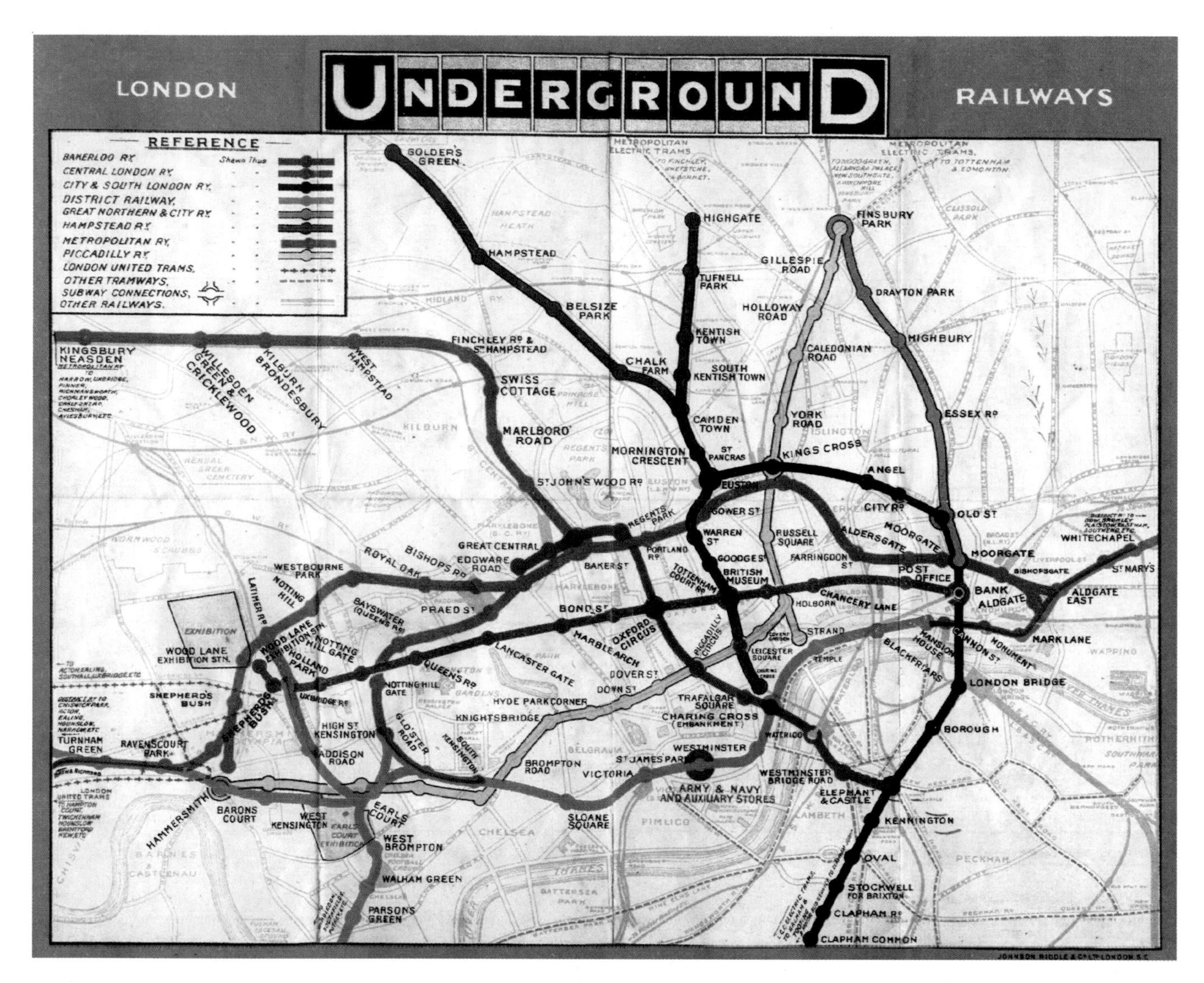

ABOVE 1908 Underground map. *(London Transport Museum)*

completed a law degree at London University the same year. He wasn't interested in pursuing a legal career so instead began working for the railways.

Through inspired determination, vision and business acumen Stanley and Pick would eventually forge London's transport system into one of world-class excellence. Stanley, after his initial misgivings, would stay for a further 40 years ending up in 1920, via a knighthood in 1914, as Baron Ashfield of Southwell, a central pillar of the British establishment.

Working together, they immediately started to get a firm grip on wayward operational procedures and began to examine various ways to increase revenue. The most obvious starting point was to avoid unnecessary duplication, and expenditure, in the design of publicity literature for the three lines, which had previously each developed their own approach to how their services should be offered to the travelling public. This was particularly noticeable in the different graphic styles and typefaces used to promote the Bakerloo, Piccadilly and Hampstead tubes, such as on station entrances.

Walter Gott, a Yorkshireman, produced the company's first maps, folders and posters. A joint booking system with coordinated fares was established throughout all of London's underground railways, including those not (yet) controlled by the UERL. The first Underground map appeared in 1908 showing these harmonised services, including the Metropolitan Railway. In just one year tremendous progress had been achieved in establishing a strong visual

ABOVE Surviving Chromo bullseye sign. *(Mark Ovenden)*

branding for the company – 'UndergrounD' – and presenting it in a consistent manner, not only on the masthead of the map but also on the posters describing destinations and services, and the fascias of stations. Stanley and Pick also introduced a common advertising policy, which improved the appearance of stations by standardising poster sizes, limiting their number and controlling their position in order to provide a considered, orderly presentation – previously passengers had complained that the uncontrolled proliferation and haphazard positioning of posters and notices made it difficult to identify the names of stations.

In addition Pick considered that a distinctively coloured surround that would make a station name panel stand out would solve this problem, and experiments were made the same year by adding two red half-moon shapes above and below the name. Where appropriate, in the older dark and dingy stations, these would also employ a white background or simply be applied as a red disc on the brighter walls of the tube stations.

These experiments were highly successful, because by the middle of the year lucrative orders were placed with a company called Chromo, enabling all UERL stations to be retrofitted with the revised name panels in a six-year programme. This manufacturer's name can still be seen today on a few surviving examples positioned on Covent Garden and Caledonian Road platforms.

Achieving financial success

Little by little the UERL's enterprising top management was able to claw its way back to solvency, and in 1908 the working losses at last turned to profits. Receipts from the District rose from £488,000 in 1906 to £691,000 by 1910, when the three tube lines were earning even more – £729,000. In this year the three UERL tube companies were officially merged to become the London Electric Railway (LER), though remaining a subsidiary of the parent organisation. Having seen his two able lieutenants starting to achieve substantial improvements, Sir George Gibb left the UERL in April 1910 to take up governmental high office, and Stanley then took on the role of managing director at the age of 35, after which he embarked on a quest to swallow up the company's rivals.

Having already got one tramway operator under its control, the London United, in 1912 Stanley snapped up another, the Metropolitan Electric. A far more significant fish netted the same year was the London General Omnibus Company, which ran most of the capital's buses. As stated before, buses had now reached a mature phase in their reliability record, and unlike the trains were also very profitable. The new group of companies, known as the London Traffic Combine, was now able to use its buses and trams as feeders rather than competitors to the Underground system. In 1913 two of the three remaining independent tube railways, the City & South London and the Central London, were added to the group, but the full-size City & Great Northern, as has already been noted, had been taken over by the Metropolitan Railway the same year.

The Metropolitan also operated the Hammersmith & City, the East London line and, for historical reasons, the top half of the Circle, which they shared with the District, who operated the lower half. An offer of amalgamation was made, also in the same year, but was firmly rebuffed by the Metropolitan, who still saw themselves as a main-line railway operator and would continue running their own independent fiefdom in splendid isolation. This situation would continue for a further 20 years, until the formation of London Transport under public ownership in 1933.

ABOVE **Train of 1913 Stock shown on the Inner Circle in later life.** *(LURS collection)*

The Metropolitan electrifies

As we have already seen, the railway had also not been slow in electrifying its tracks, and had built its own power station at Neasden for the purpose. They had decided to have motor cars with two motored bogies, two sets of traction equipment and four traction motors. By contrast, the District's motor cars had only one set of equipment, one motored bogie with two traction motors and one trailer bogie. The Metropolitan's vehicles initially weren't provided with central doors, although they were of open saloon format. Built by the Metropolitan Amalgamated Railway Carriage & Wagon Company, they were fitted with Westinghouse equipment. They were first delivered in 1904 but did not go into service on their electrified route from Baker Street to Uxbridge until January 1905. Some of the cars were delivered with gated ends and some were fully enclosed; however, it was soon found that the open ends were totally unsuitable for open-air operation, so by 1906 they had all been enclosed, staying in service until 1936. Almost identical trains were also ordered for the electrified Circle and Hammersmith & City, and these all went into service at similar dates. The East London line, however, had to wait until 31 March 1913 before it was electrified, as it hadn't been included in the electrification programme following the withdrawal of District line steam locomotives in July 1905; thus it was to lay dormant during this period. The Metropolitan's steam trains ceased operating at the end of 1906 after the Hammersmith & City's electrification.

Although the joint working of the electrified Circle line by the rival Metropolitan and District line companies would eventually operate surprisingly smoothly, the joint venture certainly got off to a very shaky start. The introduction of electric traction on the 'Inner' Circle was arranged to start on 1 July 1905, but a rainstorm flooded Hammersmith Station and a train was derailed at what is now Acton Town station (Mill Hill Park).

What happened next was the result of the total lack of engineering coordination between the two of them. The Metropolitan's electrical equipment didn't harmonise with the design of its rival, since its current-collecting shoe-gear was mounted outside the bogie frames. This allowed the positive shoes to 'float' relative to the rail on curves, but they were knocked out of their mountings as soon as they reached the other company's tracks. This was because the collector shoes of the District's vehicles were mounted on a shoe-beam suspended between the axle boxes. For the second time the Met had to bow to the District's solution and install the same current-collecting equipment. Ignominiously, the original steam locos had to be hauled back into service for a three-month period until a full electrical service could finally be provided.

The Metropolitan also bought a fleet of 20 electric locomotives that ran on their

ABOVE 'Camelback' electric locomotive en route to Aylesbury. In the middle of the rake of coaches can be seen a Pullman car (of which more later). On the right is a Great Central train from Marylebone. *(Bob Greenaway collection)*

electrified lines from Baker Street, initially from 1 November 1906 to Wembley Park, and then from 19 July 1908 to Harrow-on-the-Hill, where steam locomotives would then haul the teak carriages through to Chesham, Aylesbury and Verney Junction. These were also built by Metropolitan Amalgamated, and the first ten featured a central cab – the accompanying colour picture shows one of these in action. The second series of ten was built by British Thomson-Houston and incorporated their electro-magnetic traction control equipment. They were lighter and shorter and unlike the first batch had all the aesthetic appeal of a flying brick, courtesy of their squared-off ends.

Harrow-on-the-Hill remained the end of the electrified domain until January 1925, when the locomotive exchange point was transferred to Rickmansworth following the extension of electrified tracks. This remained the transfer point for locomotive-hauled trains until 1961, when all electric locomotives were withdrawn.

Following the UERL's acquisition of the CLR and the C&SLR tube railways in 1913, Frank Pick naturally enough soon brought them into the fold by applying his red bullseyes on to their platforms and the 'UndergrounD' logo, either with or without its red discs, to herald its stations on fascias and totem signs. These replaced the rather ugly and over-dominant graphic treatment of the word 'TUBE' that the CLR had used.

The Metropolitan meanwhile had clearly been impressed with Pick's red platform bullseyes and started to produce their own version – but using diamonds. Initially introduced with a green background on the East London line in 1912, they were progressively used with a red background from 1917 onwards to signpost their own platforms. Imitation is indeed the sincerest form of flattery! Were passengers confused when both styles met cheek to jowl at stations such as South Kensington? Certainly Pick, with his love of carefully controlled and standardised order in all things, must have been extremely exasperated by such a display.

By the time it approached its 50th birthday, leading up to the start of the Great War, the network had developed at an extraordinary rate, particularly when one considers that steam-hauled trains on the sub-surface lines lasted for 40 of them. A glance at an Underground map of the time shows a system with destinations very familiar to today's passengers:

- District line trains ran through to Wimbledon, Richmond and Ealing Broadway in the west and to Barking in the east as well as other termini at Hounslow and Uxbridge (these would in later years become Piccadilly line stations with brand new buildings).
- The Hammersmith & City went from Whitechapel to Hammersmith.
- The City & South London, with its associated

Hampstead line, went from Clapham Common to Golders Green and Highgate.

- The Piccadilly Tube ran from Finsbury Park to Hammersmith.
- The Bakerloo ran from the Elephant & Castle to Watford Junction.
- The Metropolitan line ran from Aldgate through to Chesham, Aylesbury and further points west.

District line stations were designed by their own architect, Harry Ford, in a style very much influenced by Leslie Green, but more use was made of detail along the roof and additional colours applied to the honey brown faience cladding. Baron's Court and Earl's Court are two Listed stations from 1906 that

ABOVE Baron's Court. *(Author's collection)*

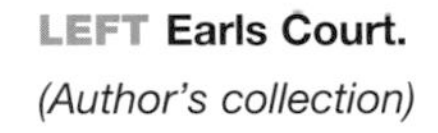

LEFT Earls Court. *(Author's collection)*

LEFT Praed Street, Paddington station. *(Author's collection)*

survive today as excellent examples of his work. Charles W. Clark had been appointed the Metropolitan Railway's chief architect, and his station entrance on Praed Street, Paddington, built in 1913–14, with its white faience cladding, pointed the way for his more grandiose stations that would be built after the war.

The new electric District line trains were a great success. They were clean, smooth and comfortable and soon became victims of their own success, because severe overcrowding became a real concern. Stanley was able to remedy this by increasing the number of trains from 24 per hour in 1907 to 40 per hour – a headway of only 90 seconds. Frequencies on the tube lines were similarly increased, with the Bakerloo operating a schedule of 34 trains per hour, while the Hampstead line running south of the major junction at Camden Town was managing a staggering 42 trains per hour by September 1909. The fact that these remarkable headways could be safely achieved was a great testimony to the inherited American signalling systems. Furthermore the last trains now left central London at 1:00am – a feature which, unfortunately, is currently unavailable to today's Underground traveller, but watch this space!.

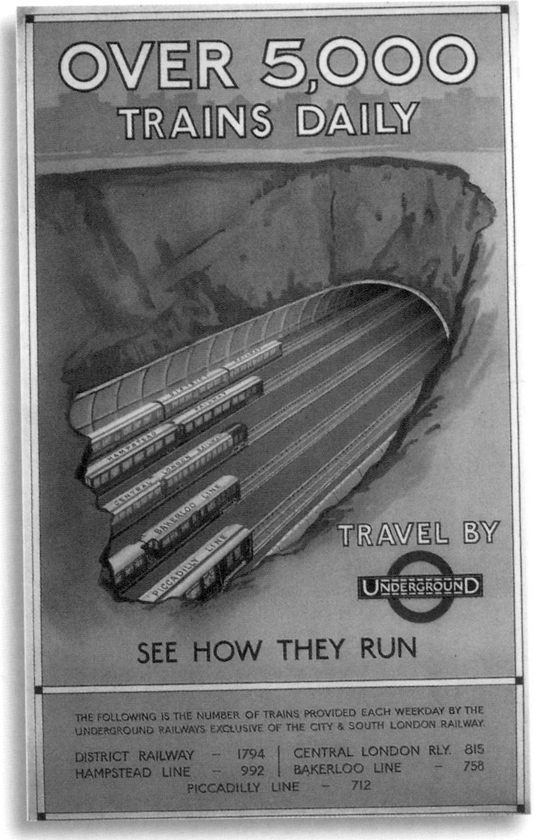

RIGHT 'See how they run' poster. *(London Transport Museum)*

But the rival Metropolitan line was not standing still either; after electrification, trains on the Hammersmith & City increased from six to eight per hour, and seven minutes were shaved from the journey time of 39 minutes from Hammersmith to Aldgate.

The Metropolitan line Pullman coaches

Metropolitan pulled off a notable marketing coup by introducing two Pullman cars on the services out to Aylesbury, having concluded an initial ten-year agreement with that company for a buffet service. They were to be the first electrically-hauled Pullman service in Europe. Introduced in June 1910, they were to compete with the opulent 'trains de luxe' first-class service provided by the Great Central Railway, which travelled on parallel tracks to the same destinations. Named *Mayflower* and *Galatea*, they were painted and lettered in the Pullman livery of dark brown and cream with gold lining that remains familiar today. Unfortunately, as might be expected, this colour scheme and tunnel dust were not a harmonious combination, so after an overhaul in 1923 they were repainted in crimson lake (a very dark red) overall, and so they remained to the end of their lives.

Luxuriously appointed with generously proportioned armchairs, with a bar at one end, they offered breakfast on the inbound service to London and a late supper on the late train returning from the West End theatres. The service lasted until October 1939, when it became one of the first casualties of the Second World War. After the conflict there was no longer a place for such extravagant creature comforts in the austere and cash-strapped 1940s and 1950s, and so the two coaches disappeared without trace.

Further line extensions were also coming thick and fast – no resting on laurels here. The main line railways were now seeing the attraction of linking with and forming extensions to the tube network, and so, partially subsidised by the Great Western Railway, the Bakerloo was extended from Edgware Road to

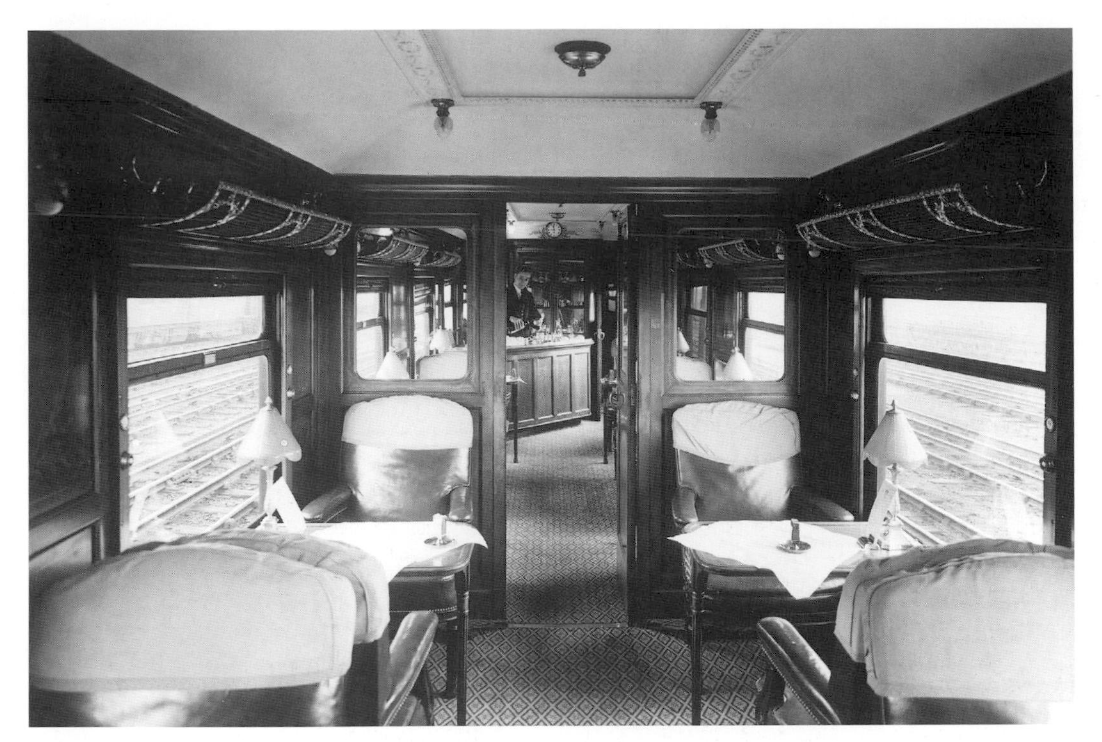

LEFT Opulent interior of one of the two Pullmans with a view through to the bar. *(London Transport Museum)*

Paddington on 1 December 1913. Escalators were provided to facilitate the steady stream of its passengers seeking the West End and other destinations. These had first been installed two years previously at Earl's Court, to link the Piccadilly and District line platforms, which is today one of the key interchanges on the Underground for trains to the five Heathrow Airport destinations.

Parliamentary approval was granted in 1912, and under an agreement with the LER the Bakerloo was to link up with the London & North Western Railway's overground electric lines at Queen's Park. A direct connection to the West End was again a very coveted destination for them, and so they agreed to fund the two-mile tube extension. Work soon began but was delayed by a shortage of labour following the outbreak of war. By stages the line reached Willesden on 10 May 1915 and then finally Watford Junction on 16 April 1917, running on the new overground electric tracks of the LNWR. Thus the Bakerloo was the first London tube to be physically connected to a main-line railway. The GWR financed the second tube railway extension for the CLR to push westwards for the 4.7-mile journey from its terminus at Wood Lane to Ealing Broadway via East Acton. Construction started in 1913 and was completed four years later, but electrification did not take place until after the war.

The rolling stock types that were ordered for these extensions are of technical and historical interest because they form the link between the

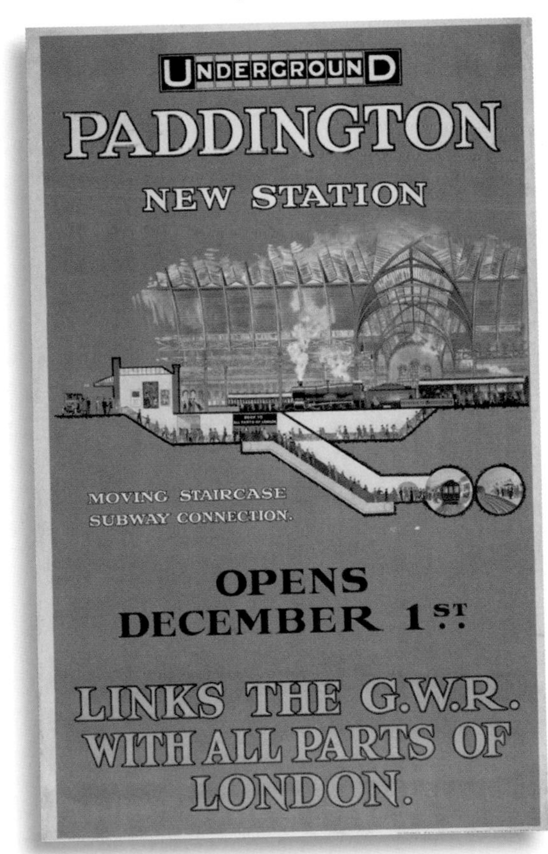

LEFT Poster showing the 1913 escalator installation at Paddington. *(London Transport Museum)*

LEFT The Leeds Forge driving motor car. *(London Transport Museum)*

Gate Stock cars and the powered doors that first appeared after the Great War.

In order to meet the service needs to Queens Park and Willesden Junction, additional trains were first obtained by transferring surplus trailer cars from the Piccadilly and running with a mixture of 14 new cars ordered in 1914; two motor cars with their trailers were built by Leeds Forge and the rest came from Brush of Loughborough. Although these vehicles still featured end gates, spring-loaded lockable centre doors that hinged inwards were now fitted on a tube car for the first time; these had red and green lights that indicated to the gateman when they were properly closed. Passenger safety and convenience was also enhanced by the routing of electric power to supply emergency battery lighting and electric tail-lamps (a first on the Underground).

The Joint Stock cars

Also in 1914 another set of rolling stock was ordered for the service out to Watford, this time from the Metropolitan Carriage, Wagon & Finance Co (later to become Metropolitan Cammell). These were known as 'Joint Stock cars' since they were co-owned by the LER and the London & North Western Railway and were painted in the latter organisation's colour scheme. Because of armaments contracts, they weren't actually built until after the war, being finally delivered in March 1920. For the first time they now had fully enclosed car ends with a similar system of interlockable spring-loaded doors, since power-operated ones were still not quite available. Another first on this stock was illuminated roller destination displays fitted on the motor cars.

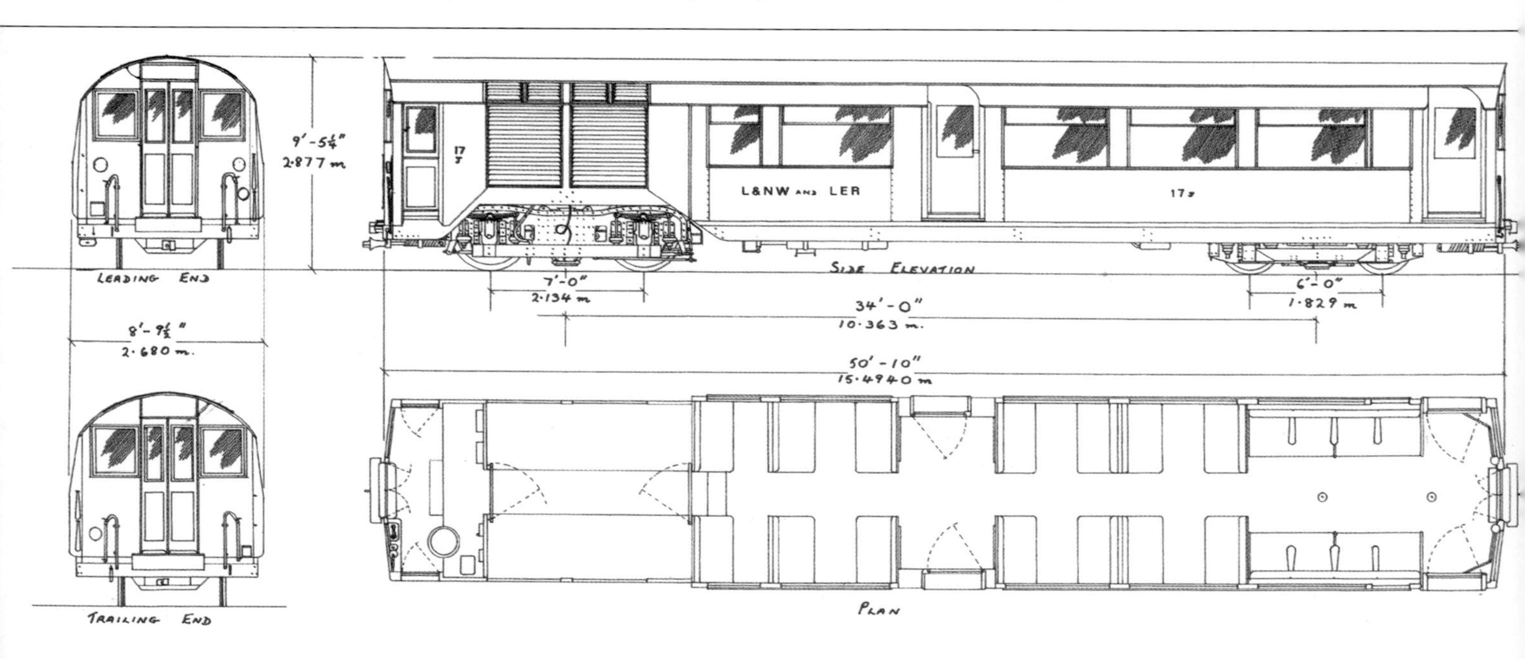

BELOW A Joint Stock driving motor. The floor height was 4.5in higher than on other Tube Stock designs, since these trains served both tube and mainline platforms beyond Queen's Park. *(Piers Connor/LURS)*

By now passengers had already been treated to fully-enclosed tube trains that were even fitted with heaters, no less. These were the vehicles of very similar design that had been ordered from Brush in 1915 for the extension to Ealing but had then to be put into storage because there were no electrified tracks for them to run on. In the event, 23 of them were lent to the Bakerloo in 1917 for their Watford services 'for the duration'. This involved them being first converted for four-rail working by the addition of outside positive shoe-gear to the motored bogie. Their successful service use, together with the dramatic increase in wartime traffic, showed that the gated ends of tube trains were no longer a viable solution and their days were now numbered. Although the manually operated 'push-in' doors on these latest train types were a step in the right direction, minds were being exercised to come up with a more convenient workable solution.

The Underground during the First World War

During the Great War the Underground experienced an exceptional growth in traffic, with 67% more passengers being carried in 1918 than at the outbreak of hostilities. This was a result of the large temporary increase in people being drawn to the capital to work on war contracts, and soldiers either travelling through London on leave or interchanging with the special troop trains on the Metropolitan line, a legacy from the very earliest days of the City Widened Lines that had been built 50 years earlier. These gave access to the overground rail network across the Thames via Farringdon, the most direct rail connection through London. An average of 16 special military trains a day travelled on these lines.

Just as in the Second World War, the deep tube system was used by Londoners as a safe haven from aerial bombardment, first by Zeppelin airships and later by four-engined bombers based on the Belgian coast. The casualty rate was thankfully minor compared to that suffered 25 years later, and the 1917 painting by Walter Bayes gives a good idea of the typical conditions of the shelterers at Elephant & Castle station.

ABOVE A 1915 Brush-built driving motor for the CLR. *(London Transport Museum)*

BELOW Elephant & Castle station in 1917. *(London Transport Museum)*

Female workers were increasingly employed to make up for the vast numbers of men being sent to the front. From the spring of 1915 onwards women were hired as guards, booking clerks, ticket collectors, lift attendants and rolling stock cleaners and painters, all of which occupations had been a male preserve prior to 1914. The photograph below shows a female 'gate-person' looking rather bemused alongside the two sets of gates on her train.

When Maida Vale was opened in 1915 – one of the stations on the extended Bakerloo line from Paddington – it was, uniquely, staffed by an all-female team. Like Kilburn Park it was designed by Stanley Heaps, who had been one of Leslie Green's assistants. Not surprisingly he therefore followed the same style. Together with the other Queen's Park extension stations it featured two more notable 'firsts' in the ongoing development of stations: the installation of its escalators was fully integrated into the station layouts, and the platforms had their station-name 'bullseyes' mounted at different heights to improve visibility for both seated and standing passengers. These treatments at Maida Vale were echoed by a magnificent large red disc mosaic 'UndergrounD' 'bullseye'l on the main entrance wall – the most pleasing and beautifully detailed example surviving on the network.

This would be the last appearance of this design format, because Frank Pick was already tinkering with a further development of it. This was during the time when he temporarily left the LER in 1917 to manage the wartime crisis in domestic fuel supplies for the government's coal mines department. His boss, Sir Albert Stanley, had similarly stepped down the previous year as the Underground's managing director to serve as President of the Board of Trade in Lloyd George's wartime government.

The Underground Group had inherited posters and information graphics that had used a variety of different typefaces, and in order to make them more authoritative and distinctive he commissioned Edward Johnston, a calligrapher and typographer, to design a unique typeface for the company in 1916. In his brief he had stated that 'it should have the bold simplicity of

RIGHT A female gate-person, who was responsible for both sets of gates. *(London Transport Museum)*

RIGHT The Maida Vale 'bullseye'. *(Mark Ovenden)*

the authentic lettering of the finest periods and belong unmistakably to the twentieth century'. Johnston's sans serif 'Underground' typeface was first used in 1916, and a slightly modified version known as 'New Johnston' is still being used today.

By 1916 Pick had decided to adapt the logo used by the LGOC, the Underground Group's bus company, which was in the form of a ring with a bar through the centre carrying the name 'General'. It is suggested that he was also influenced by the YMCA logo with its red triangle behind the word bar. The same year he commissioned Johnston to blend the new typeface on a bar against a red ring, and the world-famous 'roundel' that we know today is based on that developed by Johnston and first used in 1919.

BELOW The new 'Underground' logotype is designed. *(London Transport Museum)*

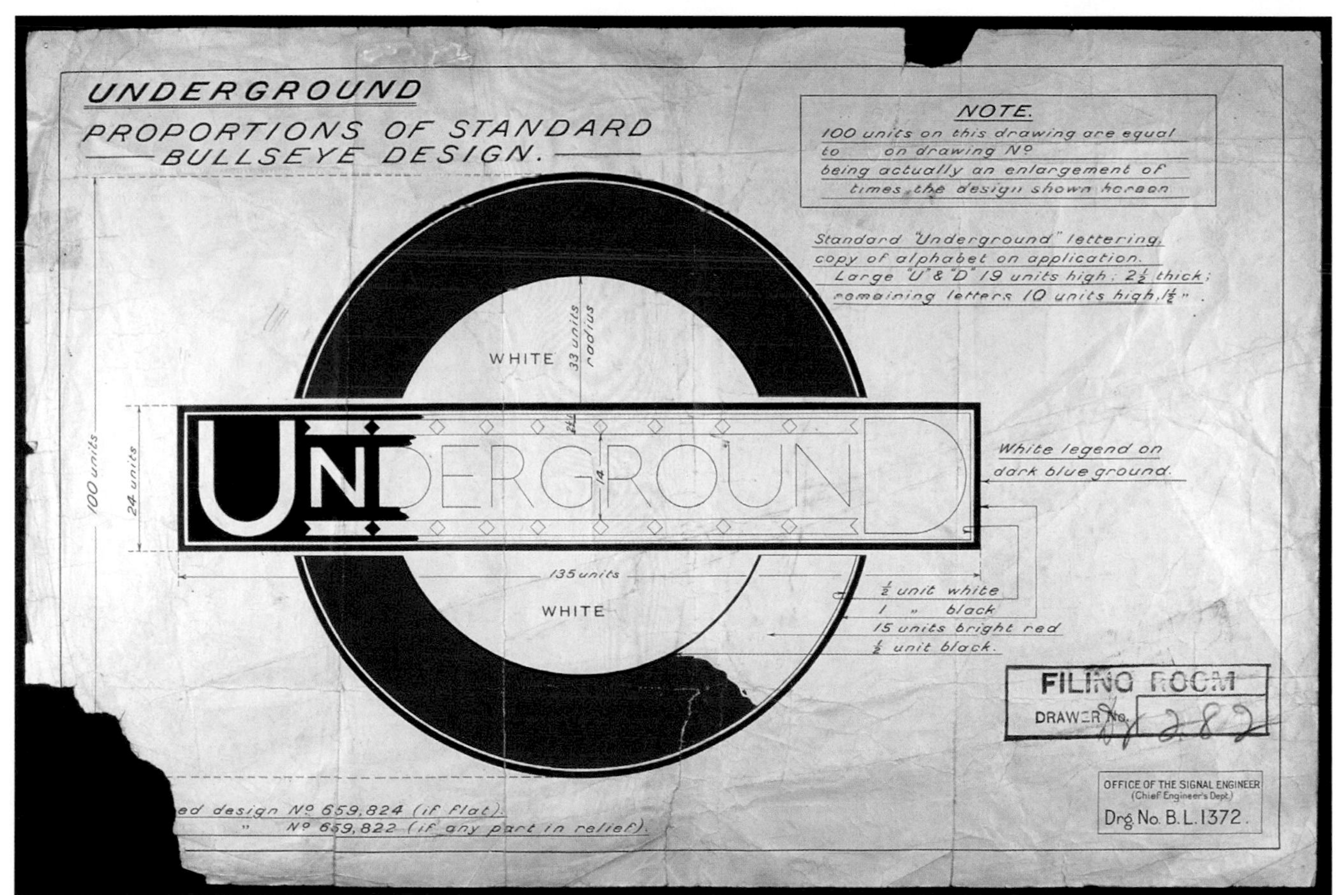

FINCHLEY CEN.
0 5 9
1044
3255

PART TWO

The second period: 1920 to 1969

OPPOSITE Two train types described in this period. The 1959/62 stock car is painted in 1924 livery to commemorate the 100th anniversary of the C&SL Railway. The other train is a 1967/72 car finished in the chosen 'corporate' livery. . *(Bob Greenaway collection)*

Chapter 5

The 1920s

The 1919/1920 Tube Stock

The first years of peace saw continual development of the system across all fronts, despite the strictures of the post-war economy; this had ushered in considerable labour unrest as the returning soldiers searched in vain for their promised 'land fit for heroes'. Symptomatic of this upheaval had been a labour strike on the Underground railway network in 1919. However, in this year – due no doubt to the tremendous technological advances made during the war – it was finally possible to put trains into service on the Piccadilly line with reliable air-operated automatic doors throughout their length.

An order was placed with Cammell-Laird for 20 control trailers and 20 trailer cars with this important feature, though the tight financial situation evidently precluded ordering matching powered motor cars to produce visually consistent train sets. To run with them, 20 old Piccadilly Gate Stock motor cars were converted to air-door operation. Two of these old French-built vehicles were later used on the Aldwych shuttle to Holborn, where they lasted until 1956.

Known as '1920 Stock', the door control system was designed to allow for two guards to operate a six-car train, thus making three 'gatemen' out of a previous total of five per train redundant. To keep the unions happy the displaced gatemen were retained as porters at their old rate of pay, to provide extra supervision on platforms; however, the national financial crisis of 1931 would see the end of this policy of employing surplus staff.

The car bodies had a pronounced bulge in their cross-section similar to the Yerkes cars, with inset doors that were nominally flat save for a curved top that echoed the roof profile. Four doors were provided down each car side, one at each end of the car and two in the middle, divided by a thick central pillar. Each

RIGHT A 1920 Stock control trailer. *(LURS collection)*

RIGHT **Poster promoting the new automatic door technology.** *(London Transport Museum)*

door leaf was provided with an 'air engine' using compressed air at an operating pressure of 30lb/in^2, and the pockets into which these doors slid were trimmed with crude and very uncomfortable upright seats. The interiors were spartan to say the least, with seats covered in a mock leather 'rexine'-type material with no armrests and a cement floor. No strap hangers were fitted. Instead they were provided with continuous horizontal handrails and vertical grab-poles that stopped short of the floor surface (to aid cleaning). Another 70 years would elapse before these features would again make their appearance (on the dramatically redesigned interiors of tube trains following the King's Cross fire in 1987). The cab of the control trailer cars featured elliptical cab windows, a solution decreed more by fashion than function, which hampered the driver's view outwards.

The design of these trains appears to have been slipped in below Frank Pick's radar, because they were the antithesis of his principle that form must follow function. With his far-sighted vision, he had stated that 'the test of the goodness of a thing is its fitness for use. If it fails on this first test, no amount of ornamentation or finish will make it any better; it will only make it more expensive, more foolish.'

His influence was soon felt, because just five years later the interiors were improved. The dreadful upright seats by the double doors were replaced by one bay of transverse seats in the centre of each half of each car, and armrests were finally provided, trimmed in the UER Group's 'lozenge' moquette fabric, as were the seats. Wooden slat flooring replaced cement, and the lighting was much improved by the addition of swan-necked lamps fitted above the window line.

These trains were transferred to the Bakerloo in 1932 because at that time the Piccadilly line was in the process of completing extensions that required a considerable amount

RIGHT **1920 Stock interior.** *(London Transport Museum)*

ABOVE The CME's instruction train. *(Bob Greenaway collection)*

of high-speed running, and the 1920 cars weren't up to the task. They were taken out of service in 1938 and later stored 'for the duration' because a plan to further recondition them for a role as a Northern & City shuttle had stalled. After the war 35 of the cars were sold for scrap, but five were retained for use as the Chief Mechanical Engineer's instruction train. Operating as a mobile classroom, this visited all of the depots with a display showing examples of equipment used on the Underground's rolling stock. Painted in light brown with black lining, the train continued in this capacity until 1968, when it was in turn scrapped. The major historical significance of such cars in the development of Underground trains was forgotten, and sadly not a single car was saved.

Improving and extending the system

The financial catalyst for post-war improvements was the government's realisation that it had to respond to the ever-increasing levels of unemployment caused by the return to peacetime conditions. In 1921 Lloyd George's coalition conceived the Trade Facilities Act, a job creation scheme whereby the Treasury guaranteed the interest on capital borrowed for new public works that would generate jobs. The capital sums still had to be found by public

subscription, but the government's guarantees made all the difference.

This was an unexpected windfall, and Ashfield and Pick took full advantage of it. The money wasn't going to be around forever, but between 1922 and 1928 the Act enabled the Underground Group to spend £12.5 million with government backing, out of a total budget of £15 million dedicated to improving the system. Fortunately the financial agreements would embrace a series of new works that had been authorised but not started before the war. So, what did the money buy?

First of all a major civil engineering programme was undertaken to reconfigure the size of the original City and South London Railway tunnels (an historic mix of diameters – 10ft 2in, 10ft 6in and 11ft 6in) to the standard LER diameter of 11ft 8in that had been established with four- instead of three-rail electrification. To give an idea of the size of the undertaking, some 22,000 tunnel-supporting rings had to be replaced. As can be imagined, this all caused major disruption during 1922–23, and a partial train service had to be abandoned due to a major (and rare) tunnel collapse near Borough Station, when 650 tons of gravel collapsed into a tunnel. It took a year before a full train service could be restored.

This project was harmonised with a new major junction between the C&SLR and the Hampstead tube at Camden Town and the extension of this line through from Golders Green to Edgware. The intricacies of this major engineering feat was the subject of a superb cutaway by the brilliant artist Leslie Ashwell Wood that appeared in a 1950 edition of the *Eagle* comic.

For now, difficult times still lay ahead, and in 1924 Lord Ashfield candidly told UER shareholders that: 'The underground railways in London have never been in their whole career a financial success. In other words they have failed to earn anything approaching a reasonable return upon the capital invested in them.'

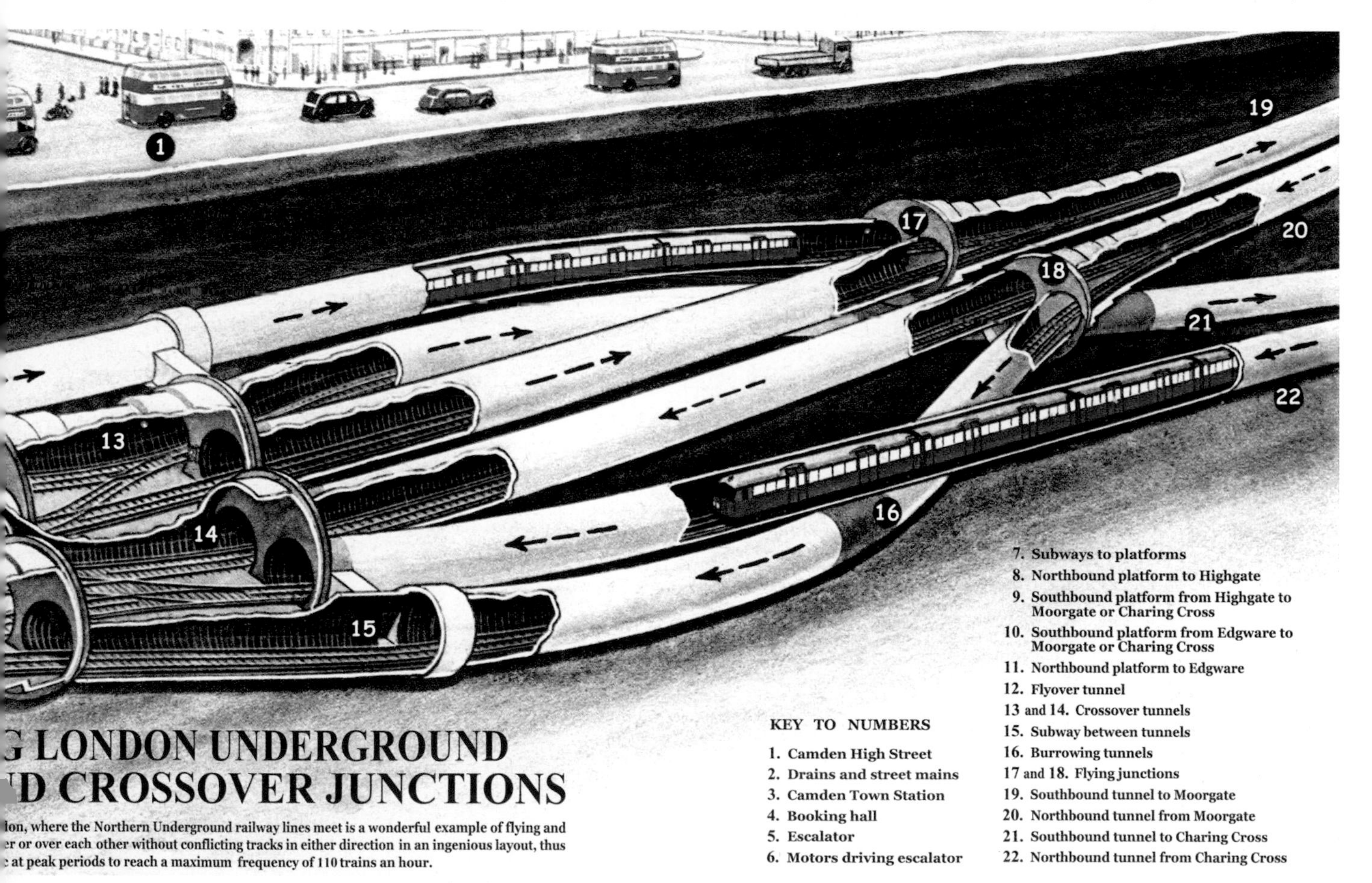

BELOW The *Eagle* cutaway. *(Reproduced by kind permission of the Dan Dare Corporation Limited)*

ABOVE Stanley Heap's design for Edgware station. *(Author's collection)*

Station design for the extension

The five open-air stations to Edgware that survive today in more or less original condition were penned by Stanley Heaps in a style that he considered suitable for reaching out into the then countryside – 'sufficiently dignified to command respect, and sufficiently pleasing to promote affection'. Apart from one – Burnt Oak, which had a simpler entrance design – they were of Georgian character, with pitched roofs and Portland stone colonnades with close-coupled Doric columns that marked the entrances. Pick was not at all sure that these architectural nods to the past expressed a modern, forward-thinking electric railway, and the Heaps designs provided the impetus for his search for a new design approach for stations as the system expanded further into the suburbs.

He turned to Charles Holden of the Adams, Holden and Pearson partnership, who he had first met in 1915 at a meeting of the Design and Industries Association (DIA). This was an organisation of which Pick had been a founder member; modelled on a German association of artists, designers and architects called the Werkbund. They had arranged an exhibition in Cologne in 1914 that was visited by a group of English like-minded individuals, who were very impressed by what they saw. (A 'model factory' at the exhibition, designed by Walter Gropius, who would later found the world-famous Bauhaus design school, featured glazed spiral staircases at both ends – a feature that Holden copied 25 years later for the new East Finchley station in 1939.)

Pick had already given Holden a 'starter project' in 1924 which was a redesigned side entrance to Westminster Station, the simple and restrained style of which foreshadowed his later Portland stone stations at the southern end of the Northern line. This entrance was demolished in 1995 to make way for the Portcullis House that was built over the new station complex for the extended Jubilee line.

Pick liked what he saw and commissioned Holden to update some of the old C&SLR stations to blend with the enlargement of their ticket halls following the installation of escalators. The exteriors were facelifted by new streamlined facades fashioned from biscuit cream faience slabs. These were relieved with

RIGHT The 1914 'Model Factory' by Walter Gropius. *(Author's collection)*

black-tiled lining and originally featured attractive 'UndergrounD' logotypes applied with individual letters in the Johnston typeface. All of these external elevations have been rebuilt over the years to a varying degree of architectural merit, and only Holden's Clapham Common station of 1924 (now Listed), with its charming domed roof and some fine original signing, remains as a surviving example of that period of his work.

The requirements of the new extensions, both existing and in the planning stage, together with the scrapping of the ancient and completely obsolete C&SLR trains, meant that ordering new vehicles was now a top priority.

ABOVE Clapham Common. *(London Transport Museum)*

The 'Standard Stock' trains make their appearance

Just as in today's practice, in 1921 a wooden mock-up of a control trailer was built at the Golders Green depot that represented the Underground Group's intentions for a new fleet of trains. Supported by a large number of engineering drawings, it was a company-owned design and specification. Oddly, despite Pick's insistence that designs had to be 'fit for purpose', the oval cab windows from the 1920 Stock with their poor outward visibility were carried over as an alternative offering, but thankfully were soon dropped.

Naturally it was by now fundamental that air-operated doors should be fitted, and in August 1922 five different manufacturers were asked to produce their solutions for five trailer cars plus one control trailer (which had to be built to rigorously detailed requirements). Although working within the framework of an overall specification, they were given a free hand to develop construction methods, interior designs and finishes, apart from observing some essential restraints. All six cars were delivered, fully finished and fitted out, a mere five months later – an astounding performance that most certainly couldn't even begin to be matched today even after the passage of over 90 years! This had a lot to do with the lack of any finite analysis in the construction methods – which were certainly over-engineered – the work-hungry character of the labour force and the depressed state of the post-war economy. However, it must be said that there is a natural limit to the number of craftsmen who can work within the confines of a tube car at any one time!

One of Pick's less successful ideas was to try to reduce the amount of noise that would be generated by these new trains, which led to all of the prototypes being fitted with shrouds over the bogies, made from materials such as horse-hide leather and pulped wood sheets. These concepts were soon abandoned, however, because they were found to be a fire hazard and a maintenance headache.

Externally the front of the driving motor cars, certainly in the first series of 1923/24, was only a minor step forward from the old Yerkes vehicles, as the accompanying photograph taken at Edgware depot in 1925 demonstrates,

BELOW The wooden mock-up at Golders Green. *(London Transport Museum)*

ABOVE Comparison between the Yerkes trains and Standard Stock. *(London Transport Museum)*

with a very similar untidy arrangement of air hoses, sockets and running lamps. These cab fronts would be tidied up on subsequent builds from 1926 onwards, and eventually all of the fleet would be modified in this way. The equipment chamber was still mounted on top of the upswept chassis frame as on all previous Tube Stock designs, but unlike their bulging passenger compartments that were visually at odds with this element, the cross-sectional form was now fully integrated with the rest of the car body. Thus a smooth, continuous contour was achieved. This seems now like an obvious step forward, but it had taken 20 years of development to accomplish it!

The interiors of the prototype cars varied from Victorian throwbacks with ornate lamp clusters and scrollwork, to some that were well ahead of their time, with smooth surfaces and concealed lighting. On all of the production motor cars built between 1923 and 1928 it was felt necessary to perpetuate the central door pillar between both sets of double doors, as it was considered that this preserved the structural integrity by helping to compensate for the massive weight at the front and back ends. On later builds it was found possible to dispense with this irritating feature that hampered ingress and egress.

By 1926 the definitive interior design was settled as illustrated, and this would become the standard solution right up to the last orders for such stock placed in 1934.

Air-operated single doors at each end of trailer cars finally became a reality in 1930, and the loss of the eight transverse seats per car was totally outweighed by the greatly increased safety and convenience for passengers, who could now exit and enter far more easily.

The poster illustration shows an earlier pre-1926 Standard Stock train at a typical Yerkes-era platform, and gives an excellent impression of the first handsome livery that was used; cars were painted in vermilion and cream with a black coach line, and all doors were picked out in maroon. They carried the 'UndergrounD' brand name in the Johnston typeface (being consistent with the lettering on Holden's facelifted stations), but car numbers were originally in an attractive but dated 'shadowed' style. However, by the middle of the decade these had also been changed to Johnston, and are today rendered in New Johnston. The cloche-hatted women's fashion

BELOW The definitive interior design with later moquette fitted. *(London Transport Museum)*

ABOVE R.T. Cooper's 1926 poster. *(Tim Demuth)*

(universal around the world) gives the scene additional period flavour. This is echoed by the similarly-hatted Marian Stanley with her father, shown alongside a special train that she had driven from Moorgate to Clapham Common in December 1924 to mark the line's extension in that year.

Since the last complete train of this type remained in Northern line service until 1966 they would come to play an intrinsic role in London's social history over four turbulent decades; and women's fashions are an interesting indicator of this, from 1920s' cloche hats and short skirts to Second World War ATS, WAAF and WRNS uniforms and their civilian utility make-do-and-mend counterparts, when women were the heroines of the 'Tube Refreshment Specials' delivering food and cups of tea to the night-time shelterers from the Blitz. These styles were followed in turn by the 'new look' of the early 1950s, and then the 'swinging London' look of the mid-1960s, and a return to short skirts.

But these cars weren't finished yet. Forty-three were then sold on to British Railways for further service on the Isle of Wight, where the youngest of them soldiered on for a further 23 years.

The Standard Stock cars would become the largest group of any rolling stock design to operate on the system, 1,446 vehicles being built over a 12-year period up to 1934. They were used on every Underground line in their time and they established the distinctive characteristics that still define today's tube trains.

LEFT Lord Ashfield and daughter Marian at the official opening of the 1924 extension to Clapham Common. *(Laurence Heyworth, Look and Learn Ltd)*

BELOW Post-1926 version of Standard Stock photographed in 1959 at Rayners Lane in 1959. The picture shows the 'cleaned-up' front end with its simple marker light cabinet integrated with the destination box. In time all of the earlier 1923–25 builds would be updated in this way. *(Ron Copson/LURS)*

LEFT The rebuilt Bond Street facade designed in 1924: the inspiration for the Morden extension stations. *(London Transport Museum)*

Tunnelling south to Morden

The Underground's major extension of the 1920s involved penetrating the newly created Southern Railway's fiercely defended electrified commuter territory. The Southern had strongly objected to the Underground Group's original plans to push south as far as Sutton, arguing with good reason that they would steal much of the business that it had created. In the end a compromise deal was struck whereby they agreed to Morden being the southernmost part of the Underground network, which it remains to this day, almost 90 years on.

Unlike the other, far cheaper, open-air extensions the five-mile extension from Clapham Common to Morden was to be in a deep tunnel, the cost being exacerbated by the fact that in that part of south London clay is sparse, water-bearing gravel beds being more usually found. Also, since the route followed under the main A24 road through Balham, Tooting and South Wimbledon considerable demolition of existing property was required, with significant compensation payable.

On 13 September 1926 the line was extended to the new terminus, and the architecture of its seven stations from Clapham South marked the transition period of Holden's work. He had moved on from his biscuit cream faience period to a new style with Portland stone frontages. This was first seen in a redesign of Harry Measure's original Bond Street exterior. Designed in 1924, but completed in 1927, this established the features that would be developed for the all-new stations – large glazed panels with vertical frames on the first floor incorporating a large 'UndergrounD' logo, a projecting canopy over large entrances, and floodlighting by night, an attractive feature that continues to this day.

Scale and full-size models were built to agree the design approach, which featured a large facade with side wings that could be arranged at varying angles to suit the

ABOVE Colliers Wood, typical example of the 'winged facade' treatment. *(London Transport Museum)*

RIGHT Evocative night shot of Tooting Broadway with its curved facade. *(London Transport Museum)*

RIGHT **Platform roundel with surround tiling.** *(Mark Ovenden)*

individual streetscape requirements. However, two stations – Tooting Broadway and South Wimbledon – differed from this concept in that they followed a curved plan, but all seven featured large roundel signs incorporated into flagpoles. Interestingly, although Pick had disliked the portico Doric columns on the stations of Heaps' Edgware extension, columns reappeared on these stations separating the first storey's three windows, but they were now fully integrated into the facade and featured an artfully contrived three-dimensional version of the roundel.

Pick had wanted to streamline and simplify the design of the stations to make them welcoming, brightly lit and efficient, with large ticket halls for the speedy sale of tickets and quick access to the trains via escalators. He was clearly pleased with the results, and at a DIA dinner in 1926 he asserted his vision, stating 'that a new style of architectural decoration will arise leading to a Modern London – modern, not garbled classic or Renaissance'. The ticket halls were fitted out in a colour scheme of off-white tiles highlighted with moulded two-tone green and purplish grey tiles that were continued down to the platforms. These framed the roundel signs that were located at two levels, as had been previously established at Kilburn Park.

This colour scheme was extensively used in many of the rebuilt Underground Group's stations around this time, and they remain very familiar to today's passengers. At Morden station the tiling around the steps leading up from the platforms to the ticket hall remain probably the best surviving examples. Westminster Station had the most extensive and well-preserved examples of this style, but they all disappeared with the building of a new station to link with the extended Jubilee line.

In time all of the stations became Listed, and during the mid-1990s they were all sympathetically restored and updated to meet today's operational requirements. The tiling and platform roundels were carefully and accurately replicated at considerable cost.

Surface and additional Tube Stock developments

The company wasn't slow in spending cash to update the fleet of trains for its District line and other Surface Stock services. The American styling influence of earlier generations of stock was soon to disappear. First came an order for 100 cars of the all-steel-bodied 'F' Stock class, ordered in 1920 from the Birmingham-based Metropolitan Carriage, Wagon and Finance Company. These were massively built, very heavy trains that were nevertheless capable of surprisingly high speeds, of which they remained capable right up to their departure in 1963 when working on the Metropolitan's service to Uxbridge. They were originally delivered with hand-worked

BELOW **'F' Stock trundling to New Cross. The appalling forward vision for the luckless driver is readily apparent.** *(LURS collection)*

ABOVE General view of an 'F' Stock train. *(Bob Greenaway collection)*

doors but were finally converted to air-door operation in 1939–40. Like the 1920 Tube Stock, the Underground Group's engineers had clearly slipped the overall design through without getting it approved by Frank Pick. The elliptical cab windows with their lack of forward visibility were certainly not 'fit for purpose', and the interiors were an ugly and unwelcoming environment with an overabundance of vertical poles, seats trimmed in 'rexine' with no armrests and a linoleum-covered concrete floor. (The Berlin Underground had a similar design operating from the mid-1920s, called the 'Tunnel Owl', and unlike London Underground has maintained a complete train still in running condition.)

BELOW Revised 'F' Stock interior. *(London Underground Museum)*

This state of affairs would never arise again. As Pick began to exercise more and more control of every detail of how the company presented itself to its public and staff, he would henceforth personally sign off in green ink all design aspects of rolling stock, stations, station hardware and publicity material for the organisation.

The poorly resolved Interiors of the 'F' Stock were meanwhile soon radically improved. Plush moquette-upholstered seats with armrests invited the passengers into an interior that was open and welcoming; the forest of vertical poles was replaced by strap hangers.

A first series of 50 'G' Stock trains were ordered in 1923 from the Gloucester Railway Carriage and Wagon Company, again with hand-worked doors, and both their interior and exterior styling details closely followed those of the 1923 Standard Stock tube cars. With no smoothing of the clerestory roof into the front cab profile, and with sheer straight sides – unlike the attractively chamfered treatment of the 'F' Stock's body – they were very 'boxy' in appearance. However, they were remarkably long-lived, the last ones remaining in service until 1971, after almost 50 years of continued service to Londoners, and one of them survives today in the London Transport Museum.

The next order was placed in 1927, with Birmingham Carriage and Wagon, for 101 'K' class vehicles, and these were certainly smoother in their appearance. As before, their front ends bore similarities to their contemporary Tube Stock cousins, with a cleaned-up

ABOVE **1923 'G' Stock.** *(H. Luff/Bob Greenaway collection)*

RIGHT **'G' Stock interior.** *(Bob Greenaway collection)*

treatment that incorporated a glazed destination display and marker light box. The details of the interior also followed those established by the tube cars from 1926 onwards.

To give another example of how Pick pursued his almost obsessive goal of train interior standardisation, the whole fleet of 1903

BELOW **1927 'K' class.** *(LURS collection)*

RIGHT Reworked Central London train with air-operated doors. *(Capital Transport collection)*

BELOW The extensive reconstruction work that was undertaken. *(Capital Transport collection)*

BOTTOM The new interior with 'lozenge' moquette; a dead-ringer for Standard Stock. *(London Transport Museum)*

Central London Railway cars were totally rebuilt in 1926 to air-door operation. The original wooden bodies received completely new car ends, new door openings, renewed hardwood framing with new steel panelling. The driving motor cars were fitted with one set of twin doors per side while the trailer cars had two sets of single-leaf doors per side. The new interiors were virtually indistinguishable to those of Standard Stock.

Although this degree of alteration was very expensive, in real terms it was the best option at the time because the new Standard Stock cars wouldn't fit into the smaller diameter of the original CLR tunnels. These heavily reworked trains remained in service until June 1939 and in the end they had to be replaced by Standard Stock. At that point in the then future the government-owned London Transport undertaking had to foot the not insignificant bill to enlarge the tunnels to take them.

Piccadilly Circus

As the 1920s drew to a close, Charles Holden pulled off his *pièce de rèsistance* for the Underground Electric Railways Company – his redesign of Leslie Green's original Piccadilly Circus station. By the early 1920s the station was becoming progressively unable to cope with the increased traffic passing through it since the fulcrum of London's centre had

ABOVE Early photograph of the Piccadilly Circus concourse. Water staining (and litter!) was clearly a problem. *(London Transport Museum)*

gradually changed from the City to the West End. People wanted to have easy access at all times of day to the ever-increasing number of cinemas, theatres and other places of entertainment as well as the fine department stores for which the surrounding areas were noted. Passenger numbers using the station tell the story: an increase from 1.5 million in 1907 to 18 million by 1922.

Stretching the limits of civil engineering technology that prevailed at the time, Holden envisaged a large elliptical subterranean ticket hall with generous circulatory areas incorporating shop-window displays that would be as large as practicable beneath the street plan of the circus itself. A large, round below street-level ticket hall at Bank, built in 1924, had already proved that the concept was feasible, so work began on the new project the same year. Costing £500,000 to build, it took over four years to complete before its official opening by Lord Ashfield at the end of 1928.

Just like the planning stages for the Morden extension stations, a full-size mock-up of the layout of ticket hall and escalator entrances, with their crossing passageways, was first constructed to test how the spaces would work. The result was nothing if not sensational, with rich, high-quality materials and finishes such as cream travertine marble, bronze fixtures and fittings highlighted by flame-red scagliola-clad columns, each highlighted by a pair of close-coupled luminaires set in a fluted bronze capital. A false ceiling was fashioned from coffered fibrous plaster panels. The lavishly detailed bronze-framed display windows, telephone kiosks, a large Mercator world map and 'see how they run' train frequency describers were all skilfully integrated into the wall surfaces.

The overall effect was initially somewhat spoiled, however, by the large tombstone-like free-standing structures which contained the automatic ticket machines. The rich finishes terminated at the concourse, the escalator shafts and passageways being tiled with the Underground Group's standard patterning of off-white with green and pale plum accents, whilst the platforms remained as originally fitted out in 1907. An intricate black and white cutaway drawing of the new station was prepared in 1928 by the artist D. Macpherson, and this was skilfully updated in colour by Gavin Dunn of the Falmouth School of Art and Design in 1989.

Holden introduced fluted bronze uplighters to illuminate the rendered cream plastered escalator shafts at this station, and these were to become a common feature throughout the 1930s. Sadly many were removed in the 1960s and only a handful of originals remain on the system, at Southgate, Swiss Cottage and St John's Wood.

Unexpectedly, the huge Moscow Underground still has similar ones, also in bronze. These were a constituent part of Stalin's grand vision of a sumptuous imperial style for his stations, with their marble walls and grandiose chandeliers (in whose reflected glory

BELOW The bronze uplighters. *(Author's collection)*

RIGHT Gavin Dunn's superb cutaway. *(London Transport Museum)*

he would bathe). When Russian engineers were researching designs for their own Underground system in the early 1930s, London was regarded as the shining example to emulate, and the new station at Piccadilly was judged to be 'the best station in London, right in the heart of the most aristocratic section'. A strange accolade indeed from diehard Communists!

In fact London Underground engineers were commissioned by the Soviets to consult on such areas as route planning, deep-tube construction methods and escalator installation. This British/Soviet collaboration came to an end in 1933 after the secret police arrested some electrical engineers for 'spying' and they were deported following a show trial. Just a year earlier Stalin had given Frank Pick a badge of merit as a token of his appreciation.

55 Broadway

As the Underground Group continued to absorb its variety of transport undertakings (tube and surface railways, buses and trams) during the first two decades of the 20th century, these were still operating from several small office buildings located near the District line's St James Park Station. By the mid-1920s, however, it had become clear that they needed to unite all of these activities under one roof, and Holden was commissioned to design a new headquarters building that would straddle the station. Its design, faced with Portland stone, was an eloquent and imposingly grand gesture that for many years was London's tallest building and demonstrated the power and grandeur of the company.

Holden hadn't previously designed anything on this scale and it seems fairly certain that he had looked to America for inspiration. He would have seen in the architectural press the development of the Cleveland Union Terminal building that was also being built above rapid-transit railway lines. Excavation of its site had started in 1922, when it was the second-largest excavation project in the world after the Panama Canal. It was built between 1926 and 1929 and was for many years the tallest building in America outside New York.

Construction of 55 Broadway commenced in 1927, was completed in 1929, and earned

LEFT 55 Broadway. *(London Transport Museum)*

Holden the RIBA London Architecture Medal in 1931. It was built on a cruciform plan that allowed the maximum amount of natural light to reach into the offices. There are certain similarities in appearance and layout to the giant Cleveland building, notably the manner in which the upper storeys step back towards a central tower that is balanced by flanking side wings. The building's interiors featured similar travertine marble facings and bronze fittings to those used in Piccadilly Circus, and an identical 'see how they run' display was also incorporated.

Totally unique was the bold commission of ten contemporary naked sculptures by a group of avant-garde artists such as Jacob Epstein, Eric Gill and Eric Aumonier. Eight of these, strongly in the art deco style, were named after the four winds and wrapped around the sixth floor pediment of the building. However, Epstein's matched pair called 'Day and Night' were placed at first-floor level more within the public's gaze, and these drew the greatest amount of public opposition. For many people, they were seen as a step too far. Indeed, on their unveiling several newspapers mounted a campaign to have them removed because they 'affronted the decency of ordinary people', and one director of the Company, Lord Colwyn, offered to pay the cost of their removal. Frank Pick, however, stood firm and even offered to resign over the absurd furore.

As he expected, the general consternation soon died down and they have remained untouched to this day (apart from some original slight surgery carried out on 'Day's' private

LEFT Metropolitan property marketing poster. *(London Transport Museum)*

ABOVE Willesden Green. *(Author's collection)*

BELOW Great Portland Street. *(Author's collection)*

parts). Having served the various forms of London Transport and London Underground for 84 years, the building is now considered no longer 'fit for purpose', and is up for sale for possible conversion into luxury flats.

The Metropolitan line during the 1920s

We have already seen how the underground railways' relentless push into the suburbs and beyond coincided with spectacular property development. The tube's Edgware extension brought with it a glut of suburban development that rapidly swallowed up the rural areas around the new stations, and this pattern was set to continue in an inexorable manner. A key factor for both railway companies during the period following the Great War was to consolidate their huge investment costs and find new ways to increase their revenue by attracting ever more passengers to use their services. Nowhere was this phenomenon more apparent than in the strategy adopted by the Metropolitan Railway, who actively marketed and promoted property development adjacent to the land that they had purchased for their own extensions into the Middlesex and Buckinghamshire countryside.

The Metropolitan's general manager Robert Selbie had been particularly pro-active in seeking new business opportunities, and in January 1919 he created a new property company, The Metropolitan Railway Country Estates Ltd (MRCE), to develop these plots. He had the vision to see that valuable revenue would come from the commuting middle classes with their season tickets, who could be enticed to buy attractive houses close to his stations, still amidst a semi-countrified environment. His posters promoted 'Metro-land' in idealistic terms as an idyllic pastoral place to live, where the fresh air of the countryside would sooth and refresh the jaded city worker, coupled with the added benefit of a fast and regular train service.

Between 1919 and 1933 – the last year of independence for the Metropolitan line – the

MRCE developed an extraordinary number of housing estates containing houses of varying style, size and quality. These were placed near their routes into the centre of London from Wembley Park, Northwick Park, Eastcote, Rayners Lane, Ruislip, Hillingdon, Pinner, Rickmansworth, Chorleywood, Amersham and Chesham. And Selbie's foresight was correct, because these locations – particularly those that remain countrified 90 years later – are particularly sought after by today's house purchasers.

From having their own power station at Neasden, through to rolling stock, station design and signing in all its forms, the Metropolitan remained resolutely independent from the rest of the Underground Group's activities and steadfastly refused to be influenced by any of their trailblazing design initiatives. In the Metropolitan's inner London locations their architect Charles W. Clark rolled out more of the same style that he had established at Paddington (Praed Street) just before the war for rebuilt stations such as Farringdon (1923), Willesden Green (1925), Edgware Road (1928) and Great Portland Street (1930), clad with his signature white faience tiling with station names rendered in red or gold serifed lettering under dentilled cornices. Of course, at this stage the Metropolitan did not have anything so distinctive and as visually powerful as the Underground's roundel, so Clark adapted the 'diamond' from their platform signs and used the device as their own marker, to discretely reinforce their corporate identity.

The diamond motif was used to highlight a station's name and also often featured as the surround to a prominent clock on the frontage. (With the passage of time only one of these now remains extant, at Willesden Green.) But with every passing year these stations appeared increasingly outmoded, and in all honesty belonged to a different era. Signing and other associated typography was also in a serifed typeface, which at least conveyed a consistent message.

But as old-fashioned as these inner London stations were, his outlying 'Metro-land' stations looked even older. They were brick built in the vernacular style of the Arts and Crafts domestic revival to harmonise with their rural surroundings. The accompanying photograph of Croxley Green station says it all. Clark completely ignored Pick's mantra that a station should serve as a distinctive beacon to the streetscape to promote the company's services. Interestingly this building style remains much loved for more expensive houses today, and is aggressively marketed as such by contemporary property developers and estate agents.

ABOVE Croxley Green. *(London Transport Museum)*

Metropolitan rolling stock

What became the most iconic trains that the Metropolitan bought in this period were the 20 1,200hp 'Bo-Bo' electric locomotives that were built by Metropolitan-Vickers at Barrow-in-Furness in 1923. (The 'Bo-Bo' terminology relates to the wheel arrangement, with four axles in two individual bogies, all driven by their own traction motors. It remains a common wheel arrangement for modern electric and diesel locomotives, as well as for driving cars in multiple-units.) These replaced the previous generation of 'steeple-backed' locos and were destined to remain in service for almost 40 years, hauling trains from Aldgate out to Rickmansworth (from 1925), where steam power took over.

They weighed 61½ tons and could reach the impressive speed of 65mph hauling varnished teak coaches. Some of these were older ex-steam-hauled stock, but they were supplemented by new coaches built between 1920 and 1923, known as 'Dreadnoughts'. They closely followed main-line practice, being of slam-door compartment configuration with luggage racks, deeply upholstered moquette

ABOVE Period coloured illustration. *(Bob Greenaway collection)*

seating and the legend 'Live in Metroland' cast into the brass door striker plates.

They made a very auspicious start, for one with its side panelling removed was proudly displayed at the British Empire Exhibition at Wembley in 1925. The venue was the Palace of Engineering, and its stablemate was the world famous 1923 LNER *Flying Scotsman* steam engine. The locomotives were originally painted in a smart livery of maroon and vermillion and all

Stewart Harris cutaway of Metropolitan 'Bo-Bo' locomotive. *(Susan Harris/Paul Ross)*

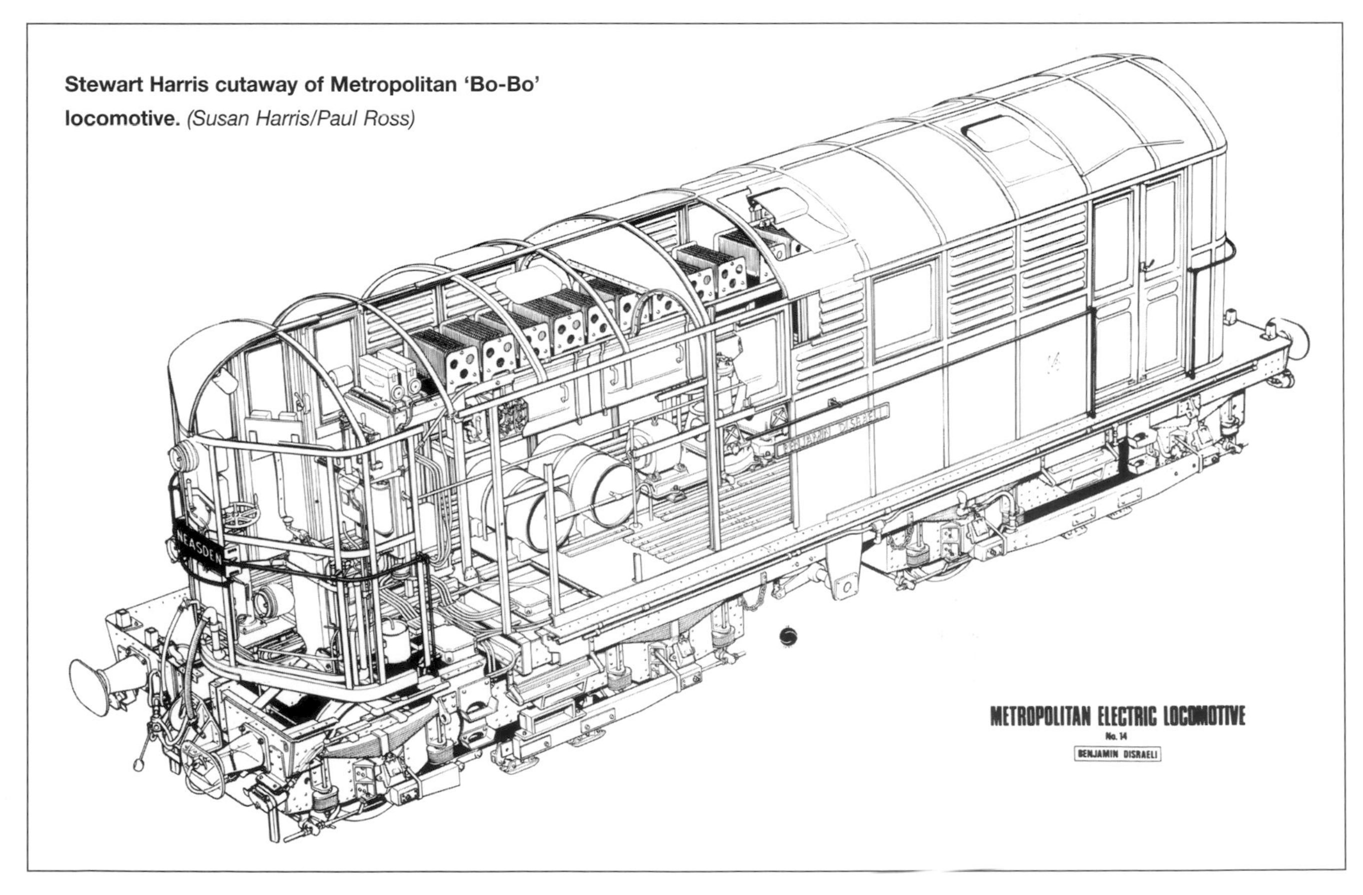

carried nameplates after prominent English men and women. During the war the locos were predictably repainted battleship grey and the bronze nameplates were removed as a token gesture to assisting the war effort (the last one wasn't taken off until 1948, three years after the end of hostilities).

In 1953, when the fleet was down to 15, they were totally renovated at Acton Works; the original colour scheme was reinstated, as were the nameplates. By 1961 they had had their day, but two survived. One of these, *John Hampden*, can be seen in the Covent Garden Museum. The other, *Sarah Siddons*, is retained at Acton for brake block testing and other duties but is regularly brought out to head enthusiasts' trains, including acting as a tender during the 150th anniversary steam run in January 2013 along the original Metropolitan tracks.

ABOVE ***Sarah Siddons*** **in fully restored condition.** *(Bob Greenaway collection)*

The Metropolitan's fleet of steam locomotives, meanwhile, was increased by a further 14 powerful new tank engines, eight 'H' class 4-4-4Ts built by Kerr, Stuart & Co being delivered in 1920–21 while six 'K' class 2-6-4Ts built by Armstrong Whitworth & Co were delivered in 1925. They were used for hauling passenger trains beyond the limits of electric traction and for the company's extensive goods services. The Metropolitan had never given up on their original ambition to be a conventional main-line railway, so they offered mixed goods trains and a door-to-door parcels service (with their own fleet of delivery vans) to augment their passenger-carrying functions.

The Metropolitan also continued to operate a variety of electric rolling stock. In the early 1920s a further order of wooden-bodied electric multiple units (virtually identical to the type ordered in 1913 for the Inner Circle and East London line) were delivered from the Metropolitan Carriage and Wagon Co. At various intervals old 'Ashbury' coaching stock was converted to electric multiple-unit drive and the earliest examples of these, with particularly large cab windows, were used on the Baker Street to Stanmore extension up to 1939.

Between 1927 and 1933 a total of 131 motor cars and trailers of a new design – initially called MV and MW (later to be known as 'T' Stock) – were built by Metropolitan Carriage and Wagon and Birmingham Carriage and Wagon. From the outset they looked very dated, with their varnished teak construction and individual slam doors opening into the same heavily upholstered interior with luggage racks etc. However, their outmoded styling concealed an important technical advance:

LEFT Metropolitan luggage collection poster. *(London Transport Museum)*

RIGHT Old coaching stock converted to multiple unit working. *(Bob Greenaway collection)*

ten of the cars delivered in 1929 featured roller bearings on the axle-boxes and traction motors. These reduced the need for regular lubrication and eliminated the problem of 'hot-boxes'. Roller bearings would be specified on all new rolling stock ordered in the 1930s and were also retro-fitted to older types. The 'T' Stock trains remained in service until 1961–62 and were withdrawn to be scrapped at the same time as the electric locomotives.

BELOW Typical 'T' Stock train. *(LURS collection)*

Chapter 6
The 1930s

In just over nine and a half years, from the beginning of the decade up to the start of the Second World War, the Underground system reached its absolute pinnacle under the ever-watchful eagle-eyed stewardship of managing director Frank Pick. It is said by those who witnessed and recorded it all, that, by August 1939, the world-beating new trains and stations, the directional signing that met the highest standards of orderliness and efficiency, and the closest attention to every detail of the functioning railway, were the outward manifestation of an organisation that was tightly run on military lines by a 'benevolent dictator' (Charles Holden's description). This was augmented by levels of housekeeping and cleanliness that have never been bettered or equalled again.

Architecture

Let us start the decade with a look at one of two stations designed by Charles Holden that can be called his transitional phase. Reconstruction of the original stations at Ealing Common and Hounslow West became necessary with the extension of Piccadilly line trains to Hounslow and South Harrow. Although work started in 1929, they were opened in March and July 1931 respectively. These were to be the last of Holden's Portland-stone-clad buildings and the essential aesthetic of the

BELOW Ealing Common. *(London Transport Museum)*

ABOVE Ealing Common station interior. *(London Transport Museum)*

'Morden' was further developed to exploit the possibilities of a 'green field' unencumbered by adjacent buildings. Both stations were very similar and featured a seven-sided 'tower' with glazed roundels set into six of the windows (the seventh being left clear, as it was behind a projecting flagpole roundel mounted on top of the station-name frieze). Holden commissioned Basil Leonides, a leading theatrical designer of the day, to develop tiled patterns for the ticket halls – a sober green, grey and cream for Ealing but a very adventurous pink, cream and flame red for Hounslow. Functional lessons had also been learnt from the Morden stations, with granite replacing Portland stone for the lower walls – more durable and resistant to staining and marking.

The Piccadilly line extensions

Proposals to extend the Piccadilly line northwards arose out of a torrent of complaints from the ratepayers of north-east London about the severe rush hour congestion at Finsbury Park. This was caused by an interchange between the main-line suburban railway, buses, trams and two tube lines that all terminated here. A Ministry of Transport enquiry in 1925 had recommended a northern extension of the line so that fewer passengers would have to break their journey here, but did not suggest where the finance would come from.

As before, an initiative from central government designed to alleviate unemployment came to the rescue. In 1929 the newly elected Labour government brought in the Development (Loans, Guarantees and Grants) Act, which offered Treasury guarantees on the interest incurred from capital loans for a period of 15 years. The Underground management once again rushed to take advantage of the scheme and quickly proposed a £13 million development that embraced far more than just the northern extension. As well as a 7.5 mile-extension from Finsbury Park to Cockfosters, the District line was widened from two to four tracks for the 4.5 miles from Hammersmith to Northfields, to enable through running of Piccadilly line trains in early 1933 to Hounslow, South Harrow and Hounslow West. Several central area stations were enlarged and remodelled, such as Leicester Square station, reconstructed between 1930 and 1935 with a new circular sub-surface ticket hall (a smaller version of Piccadilly Circus) and new escalators, claimed at the time to be the longest in the world.

With major extensions of the Piccadilly line now in the planning stages that gave the opportunity to build brand new stations, Pick felt that a fresh and distinctive style of station architecture had to be developed to herald and promote the opportunities afforded by these sites. But what should this be?

Prompted by various reviews in the architectural press showing exciting new construction projects, Pick and Holden embarked on a two-and-a-half-week tour of North Germany, Holland, Sweden and Denmark in July 1930. It is fascinating to see how the various buildings that they examined and photographed influenced the design of Holden's new stations.

In Berlin there were two new stations that clearly made a great impression, both designed by Alfred Grenander, a Swedish architect living in Germany who designed in all over 70 stations for the Berlin U Bahn.

The first was Krumme Lanke, built in 1929, which shows many architectural treatments subsequently taken up by Holden. Its sweeping curved projecting roof over the entrance with inset round light fittings, illuminated station name graphics and recessed doors (metal in

this case), all integrated between the squared entrance columns, would become very familiar themes on the new Piccadilly line stations. The columns were echoed as thick mullions between the first floor windows, and this treatment would also become familiar.

In Sweden they viewed the Stockholm Public Library building, designed by Gunnar Asplund and finished in 1928. This featured a circular brick drum, again topped by a concrete lid.

Denmark was judged as having little to offer architecturally, but Holden was impressed by the manner in which architects had also designed and had control of the installation of ancillary fixtures and fittings. This point made a great impression on him and would colour the 'look' of all of Holden's subsequent interior designs.

Finally, the new Dutch architecture they saw was much more to Holden's liking, particularly the 1928 City Hall by architect Willem Marinus Dudok that had just been completed in Hilversum. Composed of box-like forms but with a strong horizontal emphasis, the building's commanding tower at one end greatly impressed him, and he would later use this arrangement himself. Other Dutch buildings that caught his fancy were located in The Hague and Amsterdam.

Sudbury Town was the first of the new breed of Underground stations. Opened in 1931 after incredibly taking just six months to build, it replaced an old District line tin-roofed structure. It was a brick box with a flat concrete lid, pure

ABOVE Krumme Lanke station. *(DXR, Wikimedia Commons)*

LEFT The Stockholm Public Library building. *(Andreas Ribbefjord/ Swedish Wikipedia)*

BELOW Hilversum City Hall. *(nl.Wikipedia)*

ABOVE Exterior elevation of Sudbury Town. *(London Transport Museum)*

RIGHT The coffered ceiling. *(Author's collection)*

BELOW 1931 photograph showing the original neon sign. *(London Transport Museum)*

and simple, but the detailing was superb. Double-height windows were placed on both sides of the box, which creating a bright and airy ambience to an already generously proportioned ticket hall. The natural light was augmented by two daffodil-shaped uplighters (since removed and replaced by coarse, squat industrial-type fittings mounted on the base of the glazing) that illuminated an attractively coffered ceiling. This resulted in a bright, cheerful and inviting aura in the hours of darkness, helped by a deco-style illuminated neon station sign in red and blue, which survived until the early 1950s, the first and only application of neon signing on any Underground station.

At platform level the platforms received cantilevered concrete canopies, and the detailing of the rounded waiting room with its thick mullions recalls the upper storey of Krumme Lanke, as do essentially the same metal doors that give access to the waiting room. In fact the interior of this part of the building still exudes a strong Continental period flavour.

The reviews in the architectural press were most favourable. Holden's prototype station was described by historian Nikolaus Pevsner as a 'landmark' that heralded the start of a new 'classic' phase of Underground architecture. Streets of suburban houses were soon to be clustered around the new station, but Pick and Holden could never have imagined that the future would bring with it an explosion in private car ownership. Today, the heroic proportions of this and many other stations from the same period seem at odds with their empty waiting rooms and sparsely-filled, cavernous ticket halls after the morning and evening peak hours of travel have subsided. However, a visit to this station is a total time-warp experience, since the great majority of it and all its details remain exactly as first built, 83 years ago. This is due to the fact that it was Listed as a building of exceptional design merit back in 1971, the first of a whole catalogue of Underground stations to receive such recognition.

Designs for the other stations came thick and fast. Pick had been pleased with Sudbury Town but was very disappointed to find that a collection of automatic machines had been 'dumped down' on the platforms and spoilt their clean presentation. In an irritated outburst

LEFT Manor House ticket hall today, with the original ceiling and drum-shaped lights restored. Of course, the free-standing ticket office is long gone. *(Author's collection)*

that underlines his obsessive attention to detail, he stated that: 'Somehow there seems to be a desire on the part of everyone to break up and destroy the tidiness and spaciousness of this station. The only way in which the spaciousness can be filled properly is by passengers and not by impediments.'

Henceforth he issued instructions that the operating manager had to provide Holden with a list of all of the required fittings and fixtures for each new station. In this way they would be properly integrated into the whole presentation from the outset, thus avoiding any 11th-hour ad hoc additions. To reinforce this, Pick would now chair regular meetings to rubber stamp and approve every design. If some detail was unsatisfactory he would send it back to be reviewed at a later date. It is the complete integration of all of these elements that makes the Holden stations from this period so satisfying.

Work had begun on the northern and western line extensions in 1930, with the first section, from Finsbury Park to Arnos Grove, opening in 1932. The following year services were extended to Oakwood (originally named Enfield West) and finally terminated at Cockfosters in July 1933: eight stations in three years, an astounding achievement. Running in twin tunnels under established buildings, the line came to the surface just south of Arnos Grove and then continued onwards through open country except for a short tunnel working at Southgate.

The styles of all the stations were mixed and matched from a set of ingredients – all, that is, except one. Manor House was an established transport hub with bus and tram interchanges. The station's surface building was in this case particularly modest, but this was made up for by an impressive ticket hall with a distinctive ceiling decoration using circular swirls in the plaster that incorporated round lighting fittings. The first of Holden's new designs for a freestanding ticket office (or 'passimeter') was fitted, made from a bronze framework in-filled with pale green linoleum. Ticket machines and other ancillaries were neatly sited and integrated into the complete presentation, setting the standard for all of his subsequent ticket halls. All of the platforms on this section followed the German practice, which they had observed in Berlin, of being almost totally clad with tiles, in this case biscuit-cream.

This colour was relieved by a limited palette of feature colours – chrome yellow, flame red, cobalt

BELOW Original photograph of Manor House platform. *(London Transport Museum)*

RIGHT **Turnpike Lane.** *(Author's collection)*

RIGHT **Wood Green.** *(Author's collection)*

BELOW **Bounds Green.** *(Author's collection)*

BELOW RIGHT **Bounds Green lower concourse today with two surviving bronze 'daffodil' uplighters. Orange is used at this station to highlight the biscuit cream tiling.** *(Author's collection)*

blue and mid green, to highlight details such as poster panels, seat recesses and fire buckets. As an aid to ventilation air was drawn through art deco-styled bronze grilles designed by Henry Stabler. These also incorporated 'Way Out' signs, with their characteristic four-flighted arrows.

Turnpike Lane station was built as part of a shopping centre and featured a tower grafted on one side of a box. The lower escalator concourse featured a large bronze daffodil uplighter that survives today.

Wood Green had to fit a corner site within an established parade of shops, so the treatment chosen was a gently curved brick facade with a narrow band of windows.

Bounds Green had a similar tower attached at one end but this time the ticket hall was octagonal in section, this form being unique on the Underground system. Again original uplighter columns survive in the lower concourse. Because clearly one architectural practice could never have designed and detailed all of these stations, the architect Charles James worked closely with Adams, Holden and Pearson to produce the detailed designs.

Arnos Grove had been the next station to be designed after Sudbury Town, and opened in September 1932, having again been completed in a scant six months. Thought by many architectural pundits and writers to be the absolute zenith of Holden's work for the Underground, Pick nevertheless had grave misgivings about the round drum topped by a flat lid (clearly inspired by the Stockholm Library building) and only reluctantly signed-off the final drawings. However, it was judged by the *Observer* newspaper to be 'an architectural gem of unusual purity', and by all accounts Holden said that it was his favourite station. The interior was (and remains) impressively light, with eight sets of windows in the drum. The roof was supported by 16 concrete pilasters and a central column, around which was arranged a round passimeter, again of bronze and linoleum.

But whereas all of the stations thus far had been variations on a theme that betrayed certain Continental influences, the next up the line, Southgate, most certainly was not. Opened in March 1933, it was built as part of a bus/train interchange and integrated with a matching set of shopping parade buildings. The station took

ABOVE Arnos Grove today. *(Author's collection)*

BELOW Original photograph showing the linoleum-clad, bronze-framed ticket office. *(London Transport Museum)*

ABOVE Southgate by night. *(Bob Greenaway collection)*

the concept of 'roundness' to extraordinary and dramatic new heights. Particularly sensational at night, it recalls a spaceship that has just landed in the midst of suburbia; it was boldly and imaginatively detailed throughout. Original bronze uplighters survive to illuminate the escalator vault, as does another set of large bronze 'daffodil' uplighters that illuminate the lower concourse area.

Next in line was Enfield West, now known as Oakfield, which also opened in March 1933. Another station designed by Charles James under Holden's close supervision, it was essentially a scaled-up version of Sudbury Town, but on a greenfield site. It featured a vast ticket hall with a similarly coffered ceiling, and much use was made of glazed black tiles with white pointing inside the station entrance.

BELOW Lower concourse area showing both sizes of uplighters. *(Bob Greenaway collection)*

Its Listed status has ensured that the original bronze-framed passimeter has been retained (though not in use), as too has a freestanding sales kiosk. A nice touch is that the original station name graphics of white out of black have been retained, notwithstanding the fact that they don't comply with the current 'corporate' style. The station is situated some 300ft above sea level, and a plaque in the ticket hall once proclaimed that 'this is the highest point in Europe in a direct line west of the Ural Mountains in Russia'.

In hindsight, just like Sudbury Town only more so, Enfield West unfortunately was and is just too large. It's through traffic at all hours had never met Pick's most enthusiastic ambitions, because for many years building development around it was particularly slow. Amazingly he originally planned for the provision of rooms for both male and female lavatory attendants! However, its statutory Listing from 1971 onwards, together with the enthusiastic and protective attitude taken by local station managers and staff, have managed to preserve many details of this and the other fine buildings on this stretch of the line for all to enjoy. Listed protection or not, it isn't an easy task to curb the wildest excesses of official vandalism, such as electrical engineers with their ad hoc galvanised trunking and wiring indiscriminately placed without any respect for architecture.

Finally the extension's terminus at Cockfosters is reached, which opened at the end of July 1933. Unlike all of the other 'statement' station buildings on the extension (with the exception of Manor House), the road entrance was self-effacing in the extreme, consisting of low-slung horizontal planes set between two squat towers that carried flagpole 'Underground' roundels. Because the station was situated in a partial cutting, daylight would have to enter the building from above. Steps led down to a five-sided ticket hall concourse magnificently brightened by natural light, the design theme then being continued to a dramatically modern interpretation of the traditional 'train shed ' which continued over the platforms. This was manufactured throughout from reinforced concrete, the board marks of the shuttering being left exposed. This was at that time a totally new aesthetic approach that had never before been exploited in this country. The technique would be later utilised with a vengeance in new buildings from the 1951 Festival of Britain onwards.

It is a tribute to Holden's genius that this bright, airy station doesn't seem at all dated even after 80 years. Thanks to its Listed status, its visual appeal is enhanced by the retention of many fine surviving original features. These include the suspended Bauhaus-type lighting globes, original roundels and 'Way Out' signs and a large suspended clock that incorporates an illuminated platform indicator. These harmonise well with a bifurcation sign designed to current graphic standards, and modern electronic train describers.

There are two other notable stations (also Listed) from this period that betray the influence of Dutch architecture seen during Holden

TOP The exterior of Oakfield with its characteristic seating bay. *(London Transport Museum)*

ABOVE Original photograph of the interior. *(London Transport Museum)*

LEFT Cockfosters today. This design was also used later for Uxbridge. *(Author's collection)*

RIGHT Boston Manor. *(London Transport Museum)*

FAR RIGHT Osterley. *(Mark Ovenden)*

and Pick's European tour. Boston Manor, completed in early 1934, replaced a primitive wooden shack of a station from early District line days. Featuring a strong interplay of horizontal and vertical forms, its most notable attribute is a splendid art deco tower with illuminated reeded glass panels that signpost the station by night. This was a direct copy of a tower detail seen on a department store in The Hague, and today's Underground considered it significant enough to include a night shot on a set of six stamps to commemorate its 150th anniversary. Unfortunately by day it now has an incredibly ugly collection of crude railings that have been applied all over its roof surfaces in the spirit of 'health and safety gone mad'.

The other 'Dutch' station is Osterley, adjacent to the A4 Great West Road, which also opened in spring 1934. On top of a blunt square brick tower, topped by a thin concrete lid, is mounted a 70ft illuminated tower very similar to a sculpture that adorned the top of a tower of the *Telegraaf* newspaper building in Amsterdam.

Creation of the London Passenger Transport Board

Aside from the tumultuous events on the world stage (such as Adolf Hitler's rise to power), 1933 was an equally significant year for the Underground.

The coordination of all of London's public transport under a common management structure had been the subject of several proposals since the late 1920s. As we have seen, the Underground Group had achieved outstanding results under private ownership, but it was the anarchic state of the bus services that were forcing the issue because of the fierce rivalry on the streets between the Group's own London General Omnibus Company and the many independent 'pirate' operators. The chosen solution would lie in the establishment of a new public authority like the BBC and the Central Electricity Generating Board, in that it would combine public ownership with unified control. Financially it would be self-supporting and unsubsidised. The result was the London Passenger Transport Board (LPTB), formed in July 1933.

The team at the top remained unchanged. Lord Ashfield was made chairman and Frank Pick vice-chairman and chief executive. Since it was to be still dominated by these two figures, it could be seen as just an enlarged version of the Underground Group but operating as a public authority with complete autonomy over all of London's bus, tram, trolleybus and Underground routes over some 2,000 square miles. The main-line railways would not tolerate the inclusion of their own local London services, but they did agree to an arrangement to pool fare receipts and complied with the setting-up of a joint committee with the LPTB to plan future developments. This co-operation would yield some outstanding stations designs later on.

The LPTB's inherited railway network now embraced 227 route miles, and in its first year of operation carried over 415 million passengers. As far as the Underground was concerned the Metropolitan Railway was at last dragged kicking and screaming under its wing, and its pretence of being a main-line railway came to an end. It was stripped of its lightly used branch lines out in rural Buckinghamshire and in 1937 had to hand over its steam operations beyond Rickmansworth to the LNER, who inherited 18 steam locomotives as well as 270 goods wagons. Steps were also rapidly taken to bring it under the corporate umbrella, with signing and roundels consistent with the rest of the system.

The Harry Beck map

It was found that the time-honoured system of showing station's extensions laid out geographically over a map of London (as typified by F.H. Stingemore's pocket map of 1931) was becoming increasingly unclear and unhelpful. In fact the Piccadilly line extensions would prove to be literally the end of the line for such treatments. An inherent problem with such 'geographic' maps was that the centrally located stations were shown very close together and the out-of-town stations were spaced much farther apart.

Harry Beck, an engineering draughtsman working in London Underground's signals office, was aware of these ever more pressing cartographical problems. Laid off during the

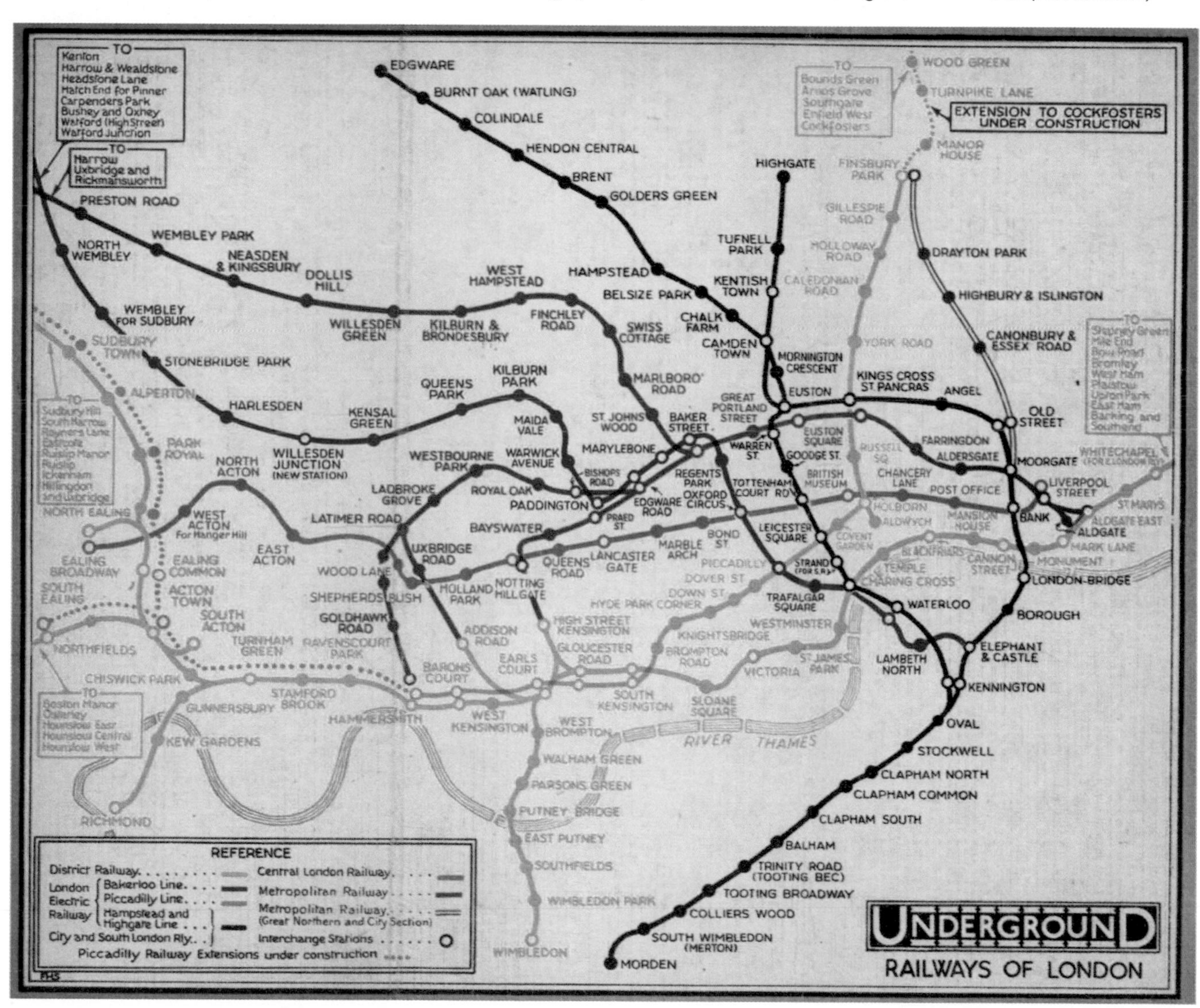

BELOW Stingemore's 1931 map. *(London Transport Museum)*

1931 Depression, he drew up a prototype route diagram inspired by his wiring diagram layouts. Without doubt the linear in-car line diagrams that had been fitted to tube cars from 1903 onwards and to Surface Stock cars from 1920, and remain essentially unchanged in their overall treatment today, were also a major influence.

The key innovation made by Beck's design was the realisation that travellers didn't need a geographic representation of the network, but rather a developed series of line diagrams showing the sequence of stations and where the interchanges were in order to reach their destination. He crucially expanded the central area and shrunk the outer areas, making the distances between all stations more or less equal – so geographical accuracy was thrown out of the window. The whole network was depicted by an arrangement of colour-coded, interconnecting horizontal, vertical and diagonal lines.

Beck first submitted his design to Frank Pick in 1931, who unpredictably considered it too radical. Not surprisingly, therefore, the publicity department also rejected it; but Beck continually pestered the management for a trial printing. Finally they relented, with an initial order for 500 folding pocket maps in 1932 – which was amazingly followed by a repeat order for 750,000 in 1933! The public had been asked for their opinion on the map and they were delighted; it was clearly a very successful design. By this time Beck was a member of staff again, working in the publicity office, but he remained a hired freelance for his map work. He was paid the mean-spirited sum of ten guineas (£10.50) for all of his initial design work, followed by an additional fee of just five guineas to painstakingly produce the master artwork for a quad royal-sized poster to be positioned in stations. In later years there was a myriad of changes and no end of graphic tinkering to the map, first by Beck (still working on a freelance basis) and later by others. Beck thought that he had absolute jurisdiction over any changes and modifications during his lifetime, but no such arrangement had ever been agreed with London Transport. He died in 1974, bitter

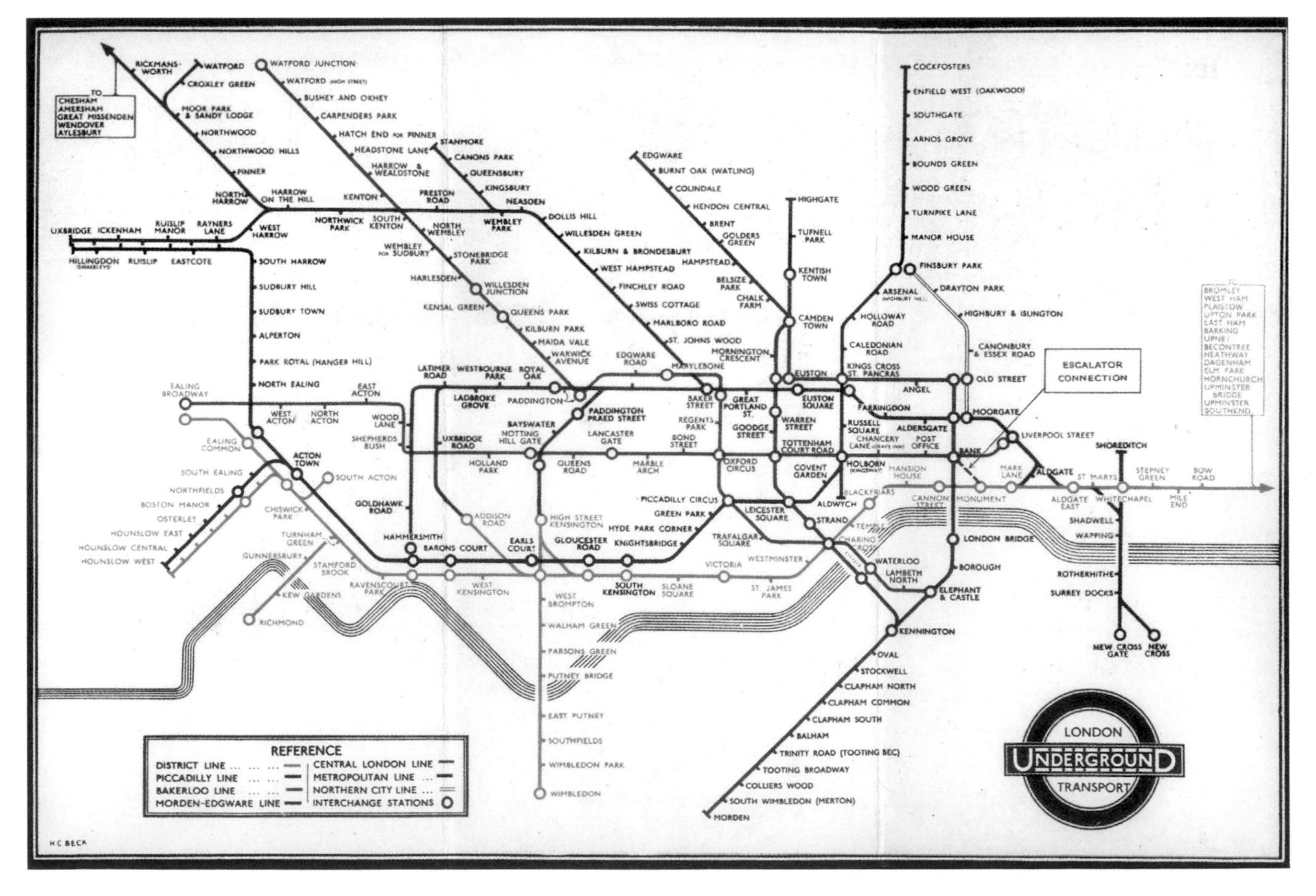

BELOW An early Beck map. *(London Transport Museum)*

to the end concerning how he and his map had been treated, but in 1997 his importance was officially though posthumously recognised and his name once more appears on every Underground map.

His lasting legacy is that virtually every metro system around the world has heavily borrowed from his design principles. The clarity of the present map, which can readily cope with every new extension and line that has to be overlaid on it, can be directly traced back to his brilliant 1933 original masterpiece. It is ironic that probably the most celebrated example of graphic design throughout the world was not produced by a trained graphic designer!

Rolling stock developments

The crisp and clean clarity of the Beck map harmonised perfectly with the Underground's new station designs, but there remained a third element of its corporate image that was still awaiting development to the same world-class standard. Although there had been further deliveries of Standard Stock tube cars in 1931 and 1934, together with a number of Surface Stock trains, the 'L' class of 1931 and the 'M' class of 1935, their designs were still from a previous era.

The 'L' class and a number of Standard Stock cars were built by the Underground Group's own manufacturing facility, the Union Construction Company in Feltham, Middlesex (where the trams and trolleybus bodies of their subsidiary London United Tramways were also built). The factory was hastily wound up following London Transport's formation, since the terms of its creation strictly prohibited the manufacture of rolling stock for its own use except for prototypes. All of the trains featured a first-class compartment located in the middle and inlaid ebonised wood trim on the inner faces of draught screens. Moquette-wool seat fabrics had now changed from the standard 'lozenge' design to patterns similar to contemporary home-furnishing fabrics.

Finally, instead of the hand-worked doors, two sets of the Metro-Cammell 'M' class trains had air-operated doors fitted with passenger-operated opening buttons; this was the first time that individual passenger door control had been fitted on any Underground train. However, they were subsequently taken out of use.

These were the last batch of trains delivered with clerestory roofs that harked back to the Underground's earliest days, and although they were fitted with innovatory smart, flush-fitting windows a new generation of stylish and smooth-running Surface Stock and Tube Stock trains was now waiting in the wings.

ABOVE A 1935 'M' class train; virtually identical to the 1931 'L' class except for powered doors being fitted. *(London Transport Museum)*

BELOW Its interior. *(London Transport Museum)*

RIGHT **William Graff-Baker.** *(London Transport Museum)*

Enter William Sebastian Graff-Baker

The Underground's American connection was by no means dead, because this man, the Chief Mechanical Engineer (Railways) to the LPTB, was born in England to American parents in 1889, and after attending various public schools went on to study at the Johns Hopkins University in Maryland, Washington DC. Following his return to England he took an electrical course at the City and Guilds College and in 1909 joined the Metropolitan District Railway as a junior fitter. His rise was meteoric; after just four years he was put in charge of all of the lifts and escalators on the LER and the CLR. In 1921 he became car superintendent in charge of all of the rolling stock depots and just one year later became assistant mechanical engineer, becoming chief mechanical engineer in 1934.

His obituary notice in 1952 stated that Graff-Baker 'had a flair for invention and design and an astonishing fertility in the exercise of these qualities which amounted to genius. He was responsible for all the major improvements in design which took place in London Transport's rolling stock, lifts and escalators such as more efficient automatic door equipment, faster and roomier trains. These all bear testimony to the ever fresh quality of mind and the greatness of his conception in design and invention.' He was by all accounts a cultured man with many interests and a deep understanding of the visual arts. Furthermore, unlike Pick (who was said to be privately a very shy man with a rather cold, aloof personality who found it difficult to praise subordinates or delegate responsibility), Graff-Baker was very affable and a fine leader of men, at the same time full of humanity and understanding.

His design philosophy was simple: 1. Will it work? 2. Is it as simple as possible? 3. Can it be easily maintained in service? 4. Can it be readily manufactured? 5. Does it look well?

In June 1935 an extensive modernisation programme was announced by London Transport in collaboration with the LNER and the GWR. This was the New Works Programme of 1935–40. The main element planned was the extension of the Central and Northern lines further into the suburbs of North London. For example, the Central line was to be extended west on new overground tracks parallel to the GWR's main line from North Acton to Ruislip and Denham. An eastern extension would run in paired tube tunnels from Liverpool Street to Stratford and then join the LNER's overground lines to Epping, Ongar and the Fairlop loop, all to be electrified for tube operation. The Northern line was to run through to High Barnet via new stations at Highgate and East Finchley, running on newly electrified LNER tracks. New Northern line extensions out to Bushy Heath and Alexandra Palace, although started, were never completed because of the war.

The New Works Programme was financed by the now time-honoured method of raising capital under Government guarantees. No public money was sanctioned but Treasury authorisation was given for the creation of a finance corporation to raise the required £40 million. Work began in 1936 on the three line extensions that were involved (Bakerloo, Central and Northern), and for these new trains were required.

The growing complex of Acton Works which the UERL had created in 1921–22 not only housed the Underground's main overhaul and repair facilities but was also the centre for all train design, development and experimental work, all headed up by Graff-Baker.

The 1935 Stock

Four six-car trains of a completely new generation were ordered from Metropolitan-Cammell in 1935 and delivered in 1937. Three of these had streamlined fronts and one had a nominally 'squared-off' flat end. The contrast between the new trains and Standard Stock could not have been greater. For a start the old noisy equipment chamber behind the driver's cab, which had been a feature of the first electric tube trains of 1903, was now replaced by a major development in traction technology that enabled all of the electrical equipment to be housed under the floor, thus freeing up space for an additional ten passengers per motor car. In fact accommodation in a six-car train of this design was now equivalent to that of a seven-car train of Standard Stock. This had been made possible by the development of a smaller type of traction motor that could be installed in each bogie. By motorising 50% of the axles the accelerating current could now approach the wheel slipping point, resulting in more power being transferred to the wheels and creating smoother and faster acceleration.

This technical breakthrough was heralded by a dramatic new styling, light years away from anything seen before on tube trains. In the mid-1930s streamlining was the outward expression of modernity, and the latest trains around the world – particularly in America – were exploiting this trend. It is therefore not really surprising that the Acton design team was seduced by these developments. The result looked more like a high-speed intercity railcar than a metro train. The rounded nose was smoothly profiled, terminating in a pronounced flare at sole-bar height. A central driving position was adopted with joystick controls and the seat could be rotated through 180° to allow passengers to exit through the front door. However, it appears that drivers didn't like this configuration not least because of inadequate draught sealing around the door.

The cab and car window line was highlighted in cream and the doors were smoothly pocketed into the body-sides. Art deco design themes influenced the feather-shaped air inlet grille on the front of the domed roof, the design of the frosted glass shovel-shaped luminaires,

LEFT The streamlined train. *(Bob Greenaway collection)*

and the seat moquette pattern of the first unit with its pink, charcoal grey and cream accents. The ceilings featured a highly polished reflective surface that maximised the light output from the light fittings. These two cars featured an early form of air-conditioning made by the Frigidaire Company at Hendon, hence the double-glazed sealed windows. These were externally flush-fitting, with rebated glass with chromium-plated inner frames. The other units had normal hinged ventilation panels. Sprung ball-ended hangers also made their first appearance, the same design being subsequently used on Underground trains for the next half century.

BELOW Its art deco interior. *(London Transport Museum)*

ABOVE The streamlined train. *(Malcolm Drabwell c/o Bob Greenaway collection)*

However, there were operational problems that came with the streamlined fronts. The large gap that was created when two units were coupled together was seen as a safety hazard – passengers could more easily fall through on to the track – and large 'bent rod' guard rails were added in an attempt to overcome this objection. Many quarters also felt that streamlining had no operational advantage given the speed of tube trains, and was judged to be an expensive fad. Although Frank Pick must have 'nodded through' the design, as seen on a scale model that was presented to management in January 1935, he is nevertheless on record as saying that he disliked the styling, which was 'not fit enough for purpose'!

However, in today's terminology they were a huge public relations success and a great image-builder for the organisation. All of the 1935 trains entered service on the Piccadilly line, where they remained until they were withdrawn in 1940. Surprisingly it is almost impossible to find pictures of the streamliners in service, but the accompanying painting of two of them at Ealing Common illustrates how they would have looked. During the war all of the 1935 Stock languished in the open air at Cockfosters depot, and in 1949 eighteen streamlined cars were simply converted to trailers. Sadly, in those days of 'make do and mend' nobody gave the slightest thought to preserving one of them for posterity.

From the original order of six trains, the flat-fronted version was in no way a less elegant solution, with its commendably clean, simple yet immaculate proportions and detailing. For example, the manner in which the domed roof, still with its 'feather' air intake, was pulled down to an arc that echoed the cab window's line was masterly. This six-car set, which was returned to service after the war and ended its days in the autumn of 1966 on the Epping to Ongar service, was the true prototype of the now iconic 1938 Tube Stock, which only differed in subtle details as seen in the photograph. For instance, the blank cream panel ahead of the first saloon window, a legacy from the streamliners, was replaced by a fourth window on production trains; in the process two extra seats per car were also gained.

BELOW The flat-fronted version. *(London Transport Museum)*

The 1938 Stock

Eventually a massive order for 1,121 vehicles of this design would be placed, shared between Metropolitan-Cammell and the Birmingham Railway Carriage Company. The adjacent colour photograph shows how the interior was finally resolved. The colour scheme of 'cerulean blue' (in reality a mid-green) combined with richly varnished woodwork, chromium-plated window frames and distinctive seating patterns, lit by warm-toned tungsten luminaires, created an ambience that was at once friendly, comfortable and welcoming. Passenger-operated door-open buttons were

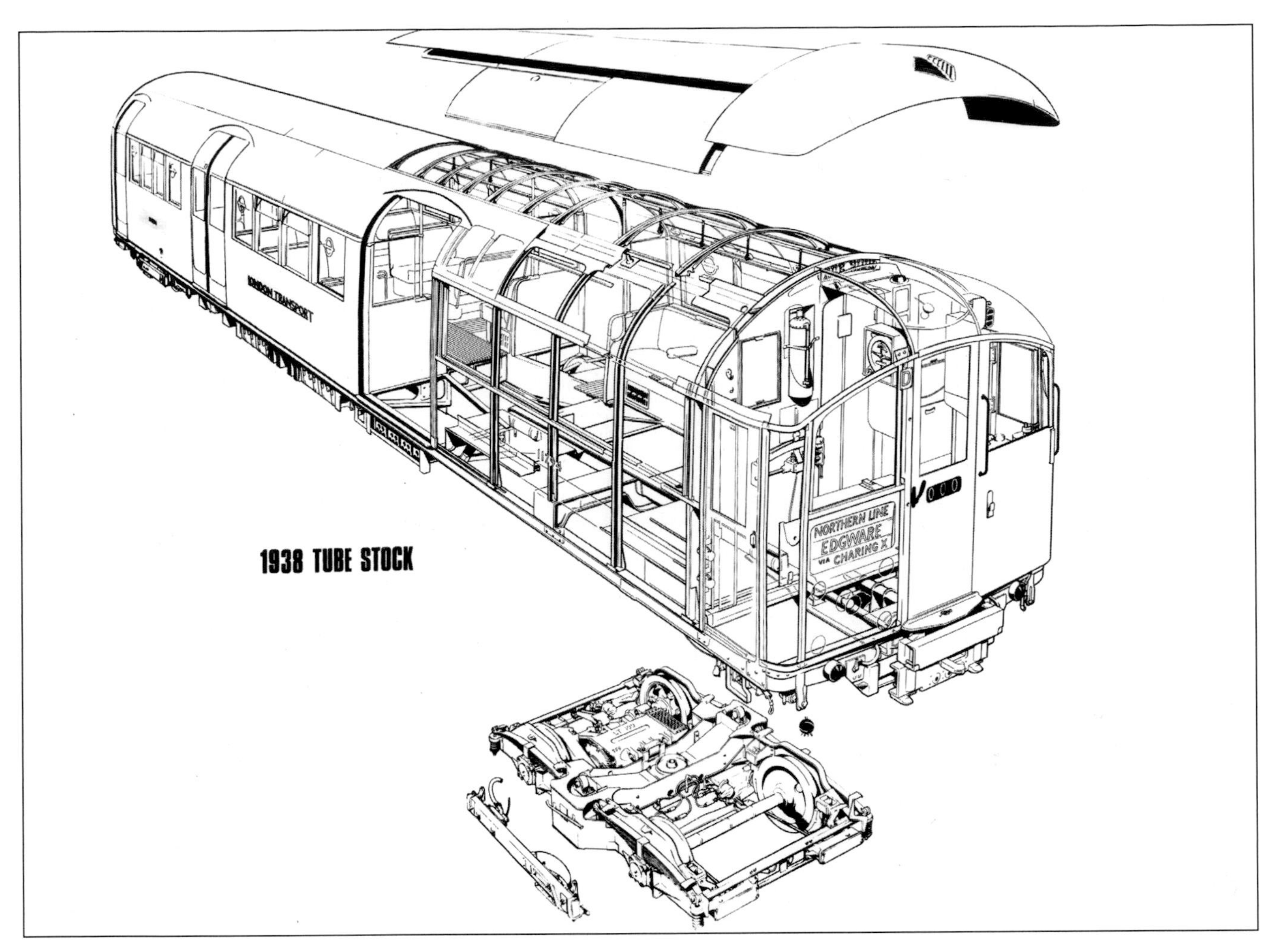

also initially fitted, but these were removed in the 1950s because guards often overrode them via a switch, causing confusion to passengers.

Some of these trains saw 50 years of operational service in London, and at the time of writing a number of them, now 75 years old, remain in service on the Isle of Wight's Island line.

New Surface Stock trains for inner suburban areas

Similar outstanding designs for Surface Stock trains also came from Graff-Baker's team at Acton in the late 1930s, culminating in the 'O', 'P' and 'Q38' types which entered service from 1937 on the Hammersmith & City line. A total of 573 vehicles were ordered from the Gloucester Railway Carriage and Wagon Co and Birmingham Railway Carriage and Wagon. These three types, externally identical, were effectively the 'big brothers' of 1938 Tube Stock. They shared a very sleek overall styling concept with rebated flush glass car-body side windows, passenger doors skilfully recessed into the car-body sides and a similar external cab treatment with a bi-coloured destination panel and round headcode lights.

ABOVE Cutaway of 1938 Stock by Stewart Harris. *(Susan Harris/ Andy Barr)*

LEFT View showing its classic interior. *(London Transport Museum)*

ABOVE 1937 'O' Stock train. *(London Transport Museum)*

RIGHT The flared-body sides compared with the previous generation. *(LURS collection)*

BELOW Interior of a Metadyne 'O' Stock car. *(London Transport Museum)*

Their most distinctive feature was the continuous flare to the lower body side, which as well as looking very stylish and distinctive had three advantages, all with safety and maintenance in mind. Firstly, they fully shrouded the door entry running boards, making it clear that it was impossible to gain a foothold when the train was moving (in the days of hand-worked doors this had always been a very dangerous temptation). Secondly, they helped to fill in the gap between train and platform. And thirdly, they assisted the mechanical cleaning of the trains.

Another thoughtful and characteristic feature of the cars' styling was the angled and chamfered treatment of the glazed panels above the windows, which stopped rain entering the ventilation panels when open.

Interior design was again closely linked to the 1935/38 Tube Stock cars with essentially the same ceiling profile, once again lit by the deco luminaires. The same relationship of ceiling contour to the fixing line of the ball-ended hangers was also maintained, but the design of draught screens (with their inlaid marquetry detailing) as well as their grab-poles was a surprising throwback to previous generations of Surface Stock cars. To aid cleaning, 45° fillets were fitted between the floor and the risers of transverse seating, another example of close attention to detail by the Acton design office.

The original seats in the 'O' Stock cars went far beyond what was normally offered on Metro trains anywhere in the world, as they were original fitted with semi-loose cushions covered with moquette that gave the characteristic deep-buttoned look of expensive upholstery. These were soon replaced (they were probably too much of a good thing) by a standard design of sprung seat. The high-backed transverse seats extending above the window line (a design also inherited from earlier types of vehicle) were only fitted on 'O' Stock motor cars; their trailer cars and all subsequent builds of 'P' and 'Q' Stock featured transverse seats that aligned with the bottoms of the windows, a convention that survives today on the newest Surface Stock vehicles.

Frank Pick had commissioned leading textile designers of their day to design unique patterns for seat fabrics that would not only look distinctive but would also wear well and disguise soiling. The designs, called 'Brent' and 'Bushey' by Enid Marx and 'Caledonian' by Marian Dorn, feature in the accompanying photograph. First-class compartments were initially located in each car, but they were withdrawn at the beginning of the Second World War and never reinstated. Passenger door-open buttons were also provided, both externally and internally, but as on 'Q38' Tube Stock they had a limited service life before being removed.

The 'O' Stock cars were fitted with a new system of electrical control developed by Metropolitan Vickers, known as the Metadyne system, together with regenerative braking. This used a rotary transformer placed between the line supply and the traction motors that required cooling air, which was ducted between two close-coupled glass panels mounted between adjacent transverse seats; on 'P' Stock the required cooling air was directed via an intake under these seats. The Metadyne units proved to be both unreliable in service and costly to maintain, and the trains were all later converted to use the pneumatic camshaft control (PCM) equipment as used on 'Q38' Tube Stock, thus becoming 'CO/CP' Stock. All three classes were finally replaced in 1981, when the last train ran in service.

More significant stations and their architects

There were many other new stations built for the Piccadilly and other lines, designed either by Holden and his practice or by other architects such as Stanley Heaps, the Underground Group's staff architect, who very ably and skilfully developed Holden's now established design principles. He is definitely an unsung hero of Underground architecture, always in the shadow of Charles Holden but quietly achieving some notable successes. Of course, way back in 1915 he had already done this for Kilburn Park and Maida Vale stations, when after Leslie Green's premature death he had to pick up the ball and run with it. This time the opportunities came from pulling down antiquated District line stations.

For example, when the Piccadilly line service took over the former District line extension to Hounslow West, a new depot was required at Northfields to stable the extra cars, and this required demolition of the old station. Another 'brick box' was therefore designed and opened in December 1932.

Sudbury Hill was yet another station with a very similar frontage that was completed in 1932–33. A fine station in the Holden mould but developed to fruition by Heaps was Chiswick Park on the District line, completed by the end of 1933. On this occasion a larger Arnos Grove-type glazed brick drum was deployed, but with a squat tower at one end bearing a large Underground roundel. In addition Heaps worked on two more variants of the

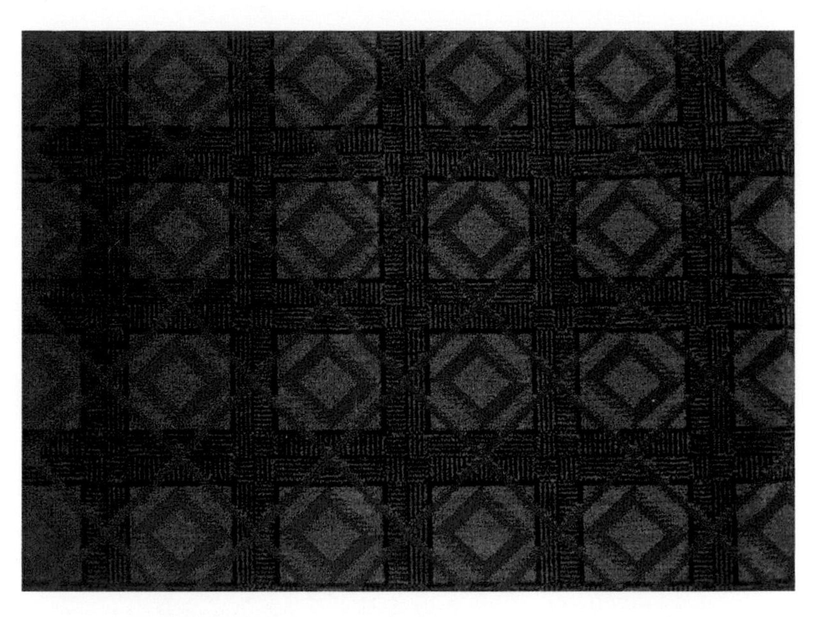

ABOVE 'Brent' moquette designed by Enid Marx. *(Bob Greenaway collection)*

BELOW Chiswick Park. *(Author's collection)*

ABOVE **St John's Wood in the early 1960s before the drum was crudely built over.** *(London Transport Museum)*

'Sudbury box' – Acton Town and Alperton, both completed in 1933 with Holden acting as consultant. Moreover he had also designed the cantilevered concrete platform canopies at Chiswick Park, Arnos Grove and Oakwood.

His legacy of fine buildings continued with the round Warwick Road entrance to Earls Court station completed in 1937, and in 1939 the new station at St John's Wood, which still retains its original bronze escalator uplighters. At platform level he introduced the beautifully detailed tiled continuous station name friezes in a biscuit-cream colour, augmented by Henry Stabler's tiled motifs that depicted London landmarks and local personalities such as Thomas Lord, founder of the nearby cricket ground.

These were the absolute high point of platform design, repeated pre-war at Swiss Cottage, Bethnal Green, Highgate and some other Central line stations after the war. A no-expense-spared recreation of the station platforms at Swiss Cottage, St John's Wood and Bethnal Green was undertaken in the late 1990s with very impressive results, thus safeguarding these fine examples of the Underground's architectural heritage.

RIGHT **Bethnal Green platform with Henry Stabler tiled motifs.** *(Wikipedia Sunil 060902)*

From around 1932 through to 1937, Charles Holden became increasingly occupied with producing designs for the University of London, culminating with Senate House. Because this naturally took up a lot of his staff's time, he began to farm out the design of some new Underground stations to office assistants and other consultants. Frank Pick was not too pleased with this turn of events, as he considered that Holden's design loyalties should be solely directed toward his Underground. But he need not have worried. Park Royal Station, designed by Herbert Welch and Felix Lander, who had worked for a period in Holden's office, was another triumph.

Completed in March 1936, the station was fully integrated into an estate of houses, flats and retail units, the entire architectural design of which was closely coordinated. A concrete-lidded brick drum over the station entrance was again deployed but this time it was dwarfed by a 67ft tower, incorporating four massive Underground roundels on each side, which provided an impressive view of the station when driving west along the A40 Western Avenue. The attractive watercolour (right) by Felix Lander shows the clerestory treatment of the staircase wells down to the platforms, a treatment used on many other Holden stations.

Completed in 1938, Rayners Lane, designed by Reginald Uren, was a particularly confident but final expression of the 'brick-box' concept with bold Underground flagpole roundels mounted on top of two streamlined curved ends.

Finally we come to East Finchley, which was opened for business on 3 July 1939. A particularly impressive building, it was the only former LNER station on the High Barnet branch to be completely rebuilt, to a design by Adams, Holden and Pearson working with architect Leonard Bucknell. It was Frank Pick's local station, as he lived a short distance away in Hampstead Garden Suburb. By an ironic twist of fate Holden at last had the opportunity to incorporate the same glazed staircase detail by Walter Gropius that he had so admired at the 1914 Werkbund exhibition a few months before the start of the First World War, and now, twenty-five years later, the Second World War was about to start in turn.

ABOVE Park Royal station watercolour. *(London Transport Museum)*

LEFT East Finchley's glazed staircase detail with Archer sculpture by Eric Aumonier. *(Mark Ovenden)*

Chapter 7

World War Two and post-war recovery

The Underground in 1939–45

During 1938 the government had appointed Frank Pick to plan the transport operations for the evacuation of London's civilians, and these were activated at the beginning of September 1939 on the outbreak of war. This was the first of several 'finest hours' for London Transport, and the tube system played a major role in ferrying 13,000 evacuees to Waterloo Station to be transferred on to main-line trains. This scene was repeated at Underground stations right across London, and in a couple of days nearly 600,000 people had been successfully evacuated.

In 1940 the then Southern Railway, which had inherited the original Waterloo & City line, managed to take delivery of a new fleet of 24 vehicles built by English Electric, again at the Dick, Kerr works in Preston (these had been ordered in 1939, but work was already too advanced for the order to be cancelled). These subsequently remained in service for 53 years with the SR and then, following nationalisation, with British Railways, until they were in due course replaced by a new London Underground design (see page 156). This particular tube railway has always been wholly within tunnels, and is isolated from any other railway connection.

BELOW **A 1940 Waterloo & City line train in later BR Network SouthEast livery.** *(LURS collection)*

Although some work on the Northern line extension continued until 1940, the most pressing tasks undertaken were to safeguard the network's safety. Tube stations in the central area and the tunnels under the Thames were protected against flooding in the event of the river, sewers and water mains being breached by bombing. This was achieved by the installation of floodgates and watertight doors at Charing Cross (Embankment) and Waterloo stations. Early precautions against bomb blasts were carried out at a number of subway entrances by sealing them off completely with ballast. Disused platforms and passageways at a number of central Piccadilly line stations were converted into emergency offices for the Board and the Railway Executive Committee (who oversaw the wartime operations of both the LPTB and the main-line operators), one of which (Down Street) was later used by Churchill's War Cabinet, within a few feet of passing trains.

Soon after the night bombing started in earnest, from 7 September 1940 onwards, the deep tube tunnels of the Underground came into their own as a safe haven for thousands of sheltering Londoners. At first this was not officially encouraged by the 'powers that be', so no special facilities were installed or provided for. Indeed, Frank Pick himself was totally opposed to tube stations being used as shelters in wartime, arguing that this would represent a total failure of planning by every responsible authority. As a consequence mayhem ruled and pompous official notices prohibiting the use of stations for sheltering were blatantly ignored. Since after a few weeks it was estimated that around 175,000 people were taking nightly refuge in 79 deep tube stations, clearly some kind of order and organisation was drastically needed, and Mr J.P. Thomas, the former general manager of the Underground, was hauled out of retirement to sort things out. In today's parlance, 'People Power' had won the day.

Trains continued to run until 10:30pm, when passengers and shelterers jostled for space on already crowded platforms. Gradually sheltering facilities became better organised, with special admission tickets, bunk beds and refreshment and medical services provided. The tickets were necessary because prior to their use a form of 'black market' exploitation had developed, whereby 'droppers', having gained access to the station earlier than the 6:30pm time when shelterers were officially allowed in, reserved spaces by laying clothing along the walls and then selling off each pitch later for up to half a crown (2s 6d, or 12.5p). The provision of chemical toilets also soon replaced the buckets behind a blanket that were initially provided.

Six 'Refreshment Special' trains were provided from November 1940 which supplied seven tons of food to the shelters nightly, delivering gallons of tea and piles of pies and buns. As can be imagined, these were immensely popular as well as being a great public relations coup. Numbered metal bunk beds were installed at 76 stations from late November 1940, with 22,000 erected by the end of the war, their fixing holes still visible today on some unmodernised and unrestored platforms.

This remarkable about-face by the government and London Transport was soon fully exploited for its propaganda value, and newsreels were shown around the country and overseas showed that smiling Londoners could now 'take it' in the security of their underground world.

But safety was not guaranteed. Between September 1940 and May 1941, 198 people were killed when tube shelters received direct hits. The worst of these was at Balham when a bomb penetrated the road surface above the platforms, killing 64 shelterers and four staff members. This was in addition to other tragedies such as the passengers killed and injured (the exact number remains unknown) when a bomb struck Sloane Square on 12 November 1940 as a fully loaded train was about to leave. It was, however, fortunate that the large pipe that still carries the Westbourne River over its platforms was not breached, because the entire Serpentine boating lake would have been emptied on to the District line. Another 56 people were killed at Bank station in January 1941 when a bomb breached the road surface above the subterranean ticket hall, causing unbelievable devastation.

ABOVE Painting of a 1938 Stock train at Trafalgar Square with shelterers bedding down for the night. *(Copyright Robin Pinnock/Rothbury Publishing)*

LEFT A Tube Refreshment Special! *(London Transport Museum)*

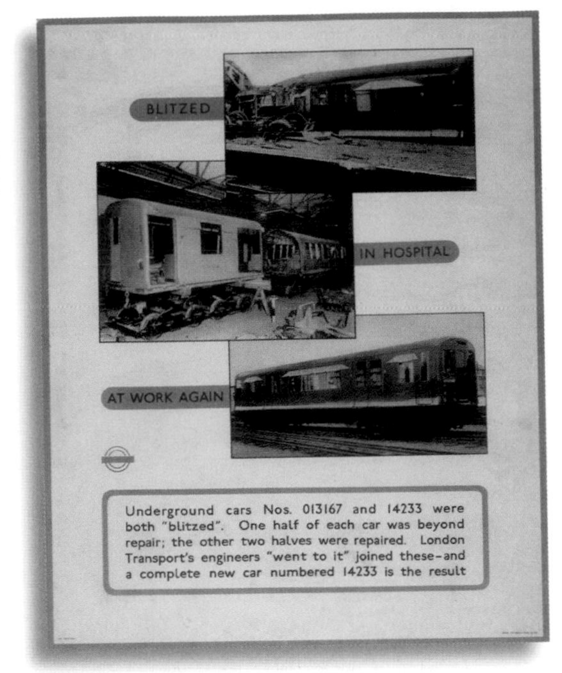

RIGHT The poster. *(London Transport Museum)*

But the worst disaster of all, not as a result of enemy action, was so appalling that the censors were able to keep it secret until after the war, when a public enquiry took place. This happened at the unfinished Bethnal Green station on the Central line in March 1943 after the local populace mistook some trial testing of anti-aircraft rockets in a nearby park for an actual air raid. They streamed down a stairwell, where a mother carrying a child tripped and fell in the dark. The ensuing crush of people falling behind her resulted in the deaths of 173 people; this death toll was the worst single civilian disaster anywhere in wartime London.

But throughout the air raids and flying bombs the Underground managed to carry on, and continued to provide a vital service. Although some sections of the system were put out of action for several months at a time, services carried on regardless. An example of the remarkable in-house skills that were still available to the organisation was the fusing together of two extensively damaged rail vehicle halves – one a motor car and the other a trailer – to form a brand new motor car. Justifiably proud of their sterling efforts, this car, numbered 14233, had the poster above left displayed in it for several years; it is one of the few of its type to survive, and is now kept by the Buckinghamshire Railway Society at Quainton Road.

Two miles of unfinished Central line extension tunnels between Leytonstone and Newbury Park were converted into an underground factory for the Plessey Company to make aircraft components. This employed 2,000

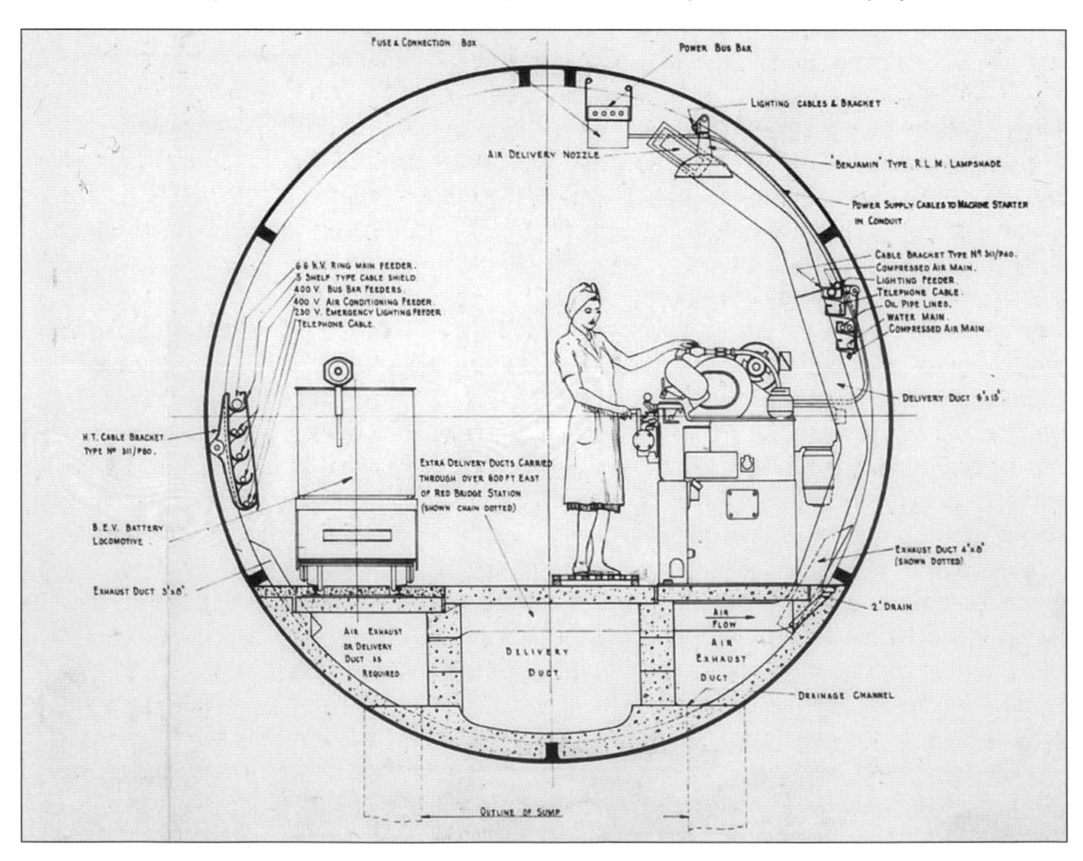

RIGHT Wartime work in an Underground tube factory. *(London Transport Museum)*

workers on day and night shifts. Acton Works overhauled landing craft engines and carried out repairs and modifications on tanks to enable them to travel in water when necessary.

Finally, the Underground's biggest contribution to the war effort was the construction of 700 Halifax bombers in the newly built railway depot at Aldenham, which was meant to stable tube cars for the Northern line extension. These were built by London Transport together with four motor companies as part of the London Aircraft Production Group.

However, as well as the vast problems involved in carrying out its essential services, London Transport had found itself unable to cover its costs towards the war's end. By 1945 fares had gone up 10% but costs had risen by 40%. For many years (in fact since the beginning of the First World War), the profitable bus network had, through a system of pooling receipts, managed to help in defraying the immense costs of operating the railway. In the forthcoming peace this situation could only get worse with the rise in private car ownership.

The wartime death of Frank Pick

Pick, who had suffered from indifferent health for many years, died of a cerebral haemorrhage at his home in Hampstead Garden Suburb in November 1941, at the comparatively early age of 62. He had retired from the LPTB in May 1940, officially on the grounds of failing health. He had previously suggested that the board's management structure be reorganised, with him continuing in a role such as joint-managing director, but this proposal was turned down. For some reason Lord Ashfield decided not to offer him such a continuing role and his position as chief executive was abolished, being replaced by a group of six heads of department.

In August 1940 he had half-heartedly accepted the position of director-general of the Ministry of Information, but he left after only four months to return to the Ministry of Transport. For them he studied improvements in the use of Britain's canals and rivers, during which time he lost two stone travelling around the country. When he died the following year he was mentally and physically worn out. His obsessiveness in getting the smallest details absolutely correct over a great number of years had finally taken its toll. Two secretaries had been continually employed by him to type out memoranda that he would sign in green ink and then fire off in all directions within the organisation. He would even fixate over such minutiae as how neatly the edge of the ballast was trimmed on the railway tracks!

Although two of his assistants would later be knighted, Pick himself had always refused such an accolade and a peerage; yet his legacy remained formidable. In 1968, Nikolaus Pevsner, the great architectural historian, wrote that: 'He was the greatest patron of the arts whom this century has so far produced in England, and indeed he was the ideal patron of our age.'

Post-war recovery – of sorts

The Labour Party's landslide victory in May 1945 was a portent of the future, and its intention to nationalise many key industries and public utilities. On 1 January 1948 the British Transport Commission (BTC) was formed to run the railways, canals and road haulage. Within this body, the London Transport Executive would replace the LPTB, and only three months earlier Lord Ashfield had announced his retirement from the board in order to join the new British Transport Commission as chairman. The appointment was short-lived, however, for he died on 4 November that year, aged 74.

Along with Pick's death eight years earlier, the demise of Lord Ashfield broke London Transport's main link with the past. These two remarkably brilliant men had, through their unique political, financial and marketing skills, masterminded the growth of the London Underground and developed it into a world-renowned railway system that employed the very highest standards of design, style and engineering in everything that it did, and provided outstanding quality of service. For sure, their likes would never be seen again.

Ashfield was succeeded as chairman by Lord Latham, the former leader of the LCC, a civil servant with scant experience of managing and operating a railway network, to shepherd

TOP Loughton station. *(Author's collection)*

ABOVE The 'Continental' ticket hall. *(Author's collection)*

BELOW One of the sculptural platform canopies. *(Author's collection)*

the Underground into what would supposedly be a bright new future.

As just one division of the BTC, London Transport was clearly not going to be the recipient of generous amounts of capital investment from the new government. Priority was given instead to other essential areas of the economy, such as house-building and electricity generation, and much of what was left in the public transport pot would go to remedy and improve the parlous state of the main-line railway system, ground down by the incessant demands of war.

Faced with the post-war reality of a Britain that was now virtually bankrupt and in hock to America, the pre-war aspirations of the New Works Programme were left in tatters. London Transport was only authorised to proceed with the unfinished eastern and western extensions of the Central line, which were completed to Epping and West Ruislip between 1946 and 1949. The new government also introduced new Green Belt planning regulations (which Pick, ironically, had supported, as he had seen how unchecked development had been greatly accelerated by new Underground lines). These meant that the proposed extension from Edgware to Bushy Heath had very little chance of developing substantial passenger traffic, so it was axed. The Aldenham depot site was therefore substantially enlarged to become a major overhaul facility for the new bus fleets being built. By this time all of the other projected extensions had been dropped, and never reappeared.

However, despite the tremendous difficulties faced at the time due to the shortages of materials of all kinds, there were nevertheless some particularly fine stations built, as well as some innovatory new trains of exceptional quality. The formation of the LPTB had opened up Underground station design to other consultancies as well as to the architects of the main-line railway companies, and they left a rich legacy. Three of the most notable designs for each section are described, most of them having particular European influences:

Eastern section

Loughton Station (Grade II listed)

John Murray Easton of the Stanley Hall, Easton and Robinson practice was commissioned by the London North Eastern Railway in 1937 to redesign all of its stations north of Woodford to Ongar, but it is thought that this station was the only one he designed. Opened by the LNER in the spring of 1940, it became part of the Underground in November 1948, when roundels were neatly added and integrated within bronze balusters. Evidently both the LNER and Frank Pick supervised the design, but the Railway company felt that although it should be in the spirit of other Underground stations, it should have its own unique style. This it has in profusion, no doubt because the LNER paid the bills!

The station is in two quite distinct halves – the building itself is distinctly North German or Dutch in character, with slim brown brickwork and recessed mortar detailing, while the ticket hall interior feels very European with its suspended Bauhaus-type spherical lamps.

The concrete platform canopies are quite unique in that they have a softly flowing sculptural quality years ahead of their time.

Gants Hill

The subterranean ticket hall accessed by subways from Eastern Avenue was opened for use in April 1948. After the first part of the Moscow Metro was opened in 1936, the Underground management had wanted to have a similar barrel-vaulted subsurface concourse in one of their stations. Charles Holden delivered the goods with this 150ft concourse with a 20ft-high vaulted roof, illuminated by fluorescent lighting. The fine quality biscuit-cream tiling was highlighted with chrome yellow bands, a particularly pleasing combination that was carried through on to the platforms with their tiled name friezes.

Western section

Hanger Lane

Twenty years after the Berlin U Bahn's seminal Krumme Lanke station of 1929, the Underground received its own very similar imitation. Designed by Peter MacIver, a British Railways Western Region architect, it was opened in January 1949. It featured a spacious ticket hall below street level (accessed by stairwells), which was clad with biscuit cream tiles up to ring-beam height A modernised, simplified yet immaculately detailed version of

ABOVE The 'Moscow Hall' subterranean concourse area at Gants Hill. *(Mark Ovendon))*

LEFT Platform clock detail. *(Mark Ovendon)*

BELOW Hanger Lane. *(Mark Ovendon)*

RIGHT **Perivale.** *(Mark Ovendon)*

BELOW **White City station.** *(Author's collection)*

Arnos Grove, it looks particularly impressive at night, but is today rather like a ship lost at sea, seemingly abandoned within the complex A40/ North Circular Road gyratory system.

Perivale (Grade II listed)
Close to the Hoover Building, in its day the largest vacuum cleaner factory in Europe, this station was designed by GWR architect Brian Lewis in 1938 but was not built until after the war, being opened in June 1947. Breaking new ground away from box and cylindrical forms, it was a breath of fresh air with its concave frontage married to a continuously undulating name frieze.

White City
This station was also designed before the war, when construction was planned to be completed by April 1940. However, built in stages, it was not finally completed for traffic until May 1950. It was planned under the supervision of the then LT chief architect Thomas Bilbow (who had replaced Stanley Heaps in 1943) by his assistants A.D. McGill and Kenneth J. Seymour. It is the innermost surface station on the western leg of the Central line.

Interestingly the exterior elevation was a nod right back to Holden's reworked 1927 Bond Street facade, especially when seen illuminated at night. His influence was equally strong in the interior design, featuring as it does his classic portfolio of harmonising materials. Beautifully detailed biscuit cream faience slabs up to door height are capped by a string of red and grey tiles. The cream tiling is beautifully offset by the grey and buff multicoloured internal brickwork, which still carries its applied lettering. The manner in which the bronze handrails articulate around the bullnosed squat entrance pillars is masterly. On the platforms, surviving roundels remain, neatly integrated within tangent-radiused wooden seating.

It remains a very fine, spacious and airy building whose design qualities won it a Festival of Britain Architectural Award in 1951 – the commemorative plaque can still be seen by the main entrance. Although it has still avoided Listed status, its fine features were restored as recently as 2008, gaining it a National Railway Heritage Award in 2009.

TOP Station entrance. *(London Transport Museum)*

ABOVE Combined roundel/seat unit. *(Author's collection)*

LEFT Close-up of Festival of Britain architectural award. *(Author's collection)*

Chapter 8
Rolling stock developments

William Graff-Baker had been seconded to the Ministry of Supply as Deputy Director General of Tank Production during the war, but following the end of the conflict he returned to continue leading the engineering teams at Acton Works Engineering Department.

His thirst for innovation remained unquenched, and two projects developed under his leadership showed that his mind was as fertile as ever.

Surface Stock trains

New trains for the District and Circle lines

The experience with duraluminium gained by London Transport whilst building fleets of Halifax bombers during the war was put to good use when it was decided to build a new fleet of Surface Stock trains. Known as 'R' Stock, they were built to replace the old hand-operated-door trains on the Circle and District lines.

Although the first batch (R47) was delivered with painted steel bodies, the underframes and bodies of the second consignment (R49) were made of corrosion-resistant aluminium alloy. They were virtually identical to the pre-war 'O' and 'P' types, essential differences being two large windows in the central saloon rather than four and the number of cabs required (two only, one at each end) in a six-car formation. It was the post-war shortage of steel that dictated the change in material, but the lower body weight of the car and the attendant increase in energy efficiency were attractive bonuses. Leaving them unpainted had added further appeal, as it was estimated that leaving cars unpainted saved two tons in weight on an eight-car train, this being in addition to the 5.4-ton weight reduction per car (12%) as a result

RIGHT The aluminium-bodied car by the Dome of Discovery and the uncompleted 'Skylon' on its way to the Transportation Pavilion. *(London Transport Museum)*

of the change from steel to aluminium.

One of the aluminium vehicles, left partially unpainted, was shown in the Transportation Pavilion at the 1951 Festival of Britain site at London's South Bank.

Performance trials were successful and the first unpainted train entered service in January 1953. Because its appearance was something of a culture shock, management tinkered with two different applications of a coloured red band that ran all around the train. Fluorescent lighting (which had been initially trialled in 1944) replaced the shaded tungsten bulbs and a form of rubber suspension was also introduced.

Surprisingly in those bleak times, both money and the requisite skills could still be found to equip the interiors to a high-quality pre-war standard. The same marquetry and ebonised timber inlays on the backs of both transverse seats and the inner faces of the draught screens were used. A nicely resolved 'fit for purpose' detail that aided both cleaning and comfort was the hollowed-out window panels below the window pillars in the car centre.

The last of these flare-sided trains remained in service until 1981–82.

New trains for the Metropolitan line

As the war was drawing to an end, the engineers at Acton who had remained behind with their 'reserved occupation' status intact began to concentrate their minds on the need to design a new breed of multiple-unit trains. These would ultimately replace the old teak carriage trains, be they electric locomotive-hauled or 'T' Stock multiple units, and would also operate on the planned electrified routes from Rickmansworth out to Amersham and Chesham. Fleets of saloon cars with air-operated doors, fewer seats and more standing accommodation would now be the order of the day. The most important part of the specification was to provide through car gangways so that passengers could be speedily evacuated in an emergency, especially in tunnel sections of the line, slam-door compartment stock being clearly very unsatisfactory in this respect.

The development route to a final design featured some bizarre ideas, as illustrated by prototype car No 17000 that ran in service for three years from the beginning of 1946. It was originally built with island seating arranged in sets of three, flanked by corridors on either side. The only window seats were at the car ends, and seated passengers in the centre bays could not see out of the windows when the corridors were filled by standing passengers. Moreover, the sliding doors opened directly on to seated passengers, with no weather protection, ensuring that they got a cold blast of freezing air in the post-war icy winter of 1947 (the coldest on record). One wonders what Frank Pick would have made of it!

Not surprisingly this hapless interior was very unpopular with passengers, but it was three years before it was stripped out and replaced by a layout featuring an offset corridor than ran between two seats on one side and three on

ABOVE An 'R' Stock interior with fluorescent lighting. *(London Transport Museum)*

BELOW Prototype Metropolitan car; very chilly in winter! *(London Transport Museum)*

ABOVE Sanity prevails! *(London Transport Museum)*

BELOW Prototype car renumbered 17001 with this revised interior. *(London Transport Museum)*

BOTTOM A production 'A' Stock train. *(London Transport Museum)*

the other. As before, fluorescent lighting and luggage racks were fitted. This proved to be the definitive layout and the car then ran until 1955, its job done.

The details being further refined, an order was placed with Cravens of Sheffield for 464 cars in 1959, 216 of which were to replace the venerable 'F' Stock of 1920 on the Uxbridge service, with its oval cab windows. Launched into service in 1961, the fleet was called 'A' Stock. They were exceptionally solid, well-built and handsome vehicles and remained in service for over half a century, being finally replaced in 2012. As Cravens were absorbed into the Metro-Cammell empire back in 1966, this was an extraordinary achievement.

The trains continued to operate on the whole Metropolitan line route as well as the East London line, and could run at high speeds on the open sections right to the end of their lives. When built they had a top speed of 70mph (the world's fastest in four-rail operation). By the early 2000s they were restricted to 50mph to improve reliability, but during a farewell tour in September 2012 an eight-car train was clocked at 74mph! The unsung engineering teams at Neasden Depot had always to perform remarkable feats of 'creative' maintenance to keep the trains' reliability levels up in the face of onerous service requirements.

At 116in (2,900mm), these were London Underground's widest trains, and they were fabricated from a mix of three types of aluminium – sheet, extruded and cast – all left in their natural finish, which aged differently over the

years resulting in a patchwork effect in their appearance. Interiors were solidly fashioned, with much use being made of laminated plastic and fibreglass, a novel material for Underground rolling stock at that time. Large one-piece fibreglass mouldings were cleverly designed to perform several functions at once. They formed seatbacks and draught screens in the vestibules, and also provided the transverse ceiling panels to which the grab-poles were anchored. They also carried diagrams of the line, which were easily read, as all-transverse seating was fitted. The moquette seat pattern was a blend of grey, black and red that harmonised well with the interior colour scheme. Individual aluminium luggage racks with coat hooks were also fitted (which were sold to enthusiasts at eye-watering prices by the London Transport Museum following scrapping of the fleet in 2012).

ABOVE The production interior when new. *(London Transport Museum)*

LEFT Interior of the 'Sunshine Car'. *(London Transport Museum)*

Tube Stock trains

1951/52 Tube Stock

One hundred brand new, seven-car tube trains to be known as '1951 Stock', based on the 1938 Tube Stock pattern, were being planned for the Piccadilly line to replace its own fleet of Standard Stock cars. Graff-Baker was anxious to exploit some new ideas in these, in particular by improving the outward visibility for crush-loaded standing passengers, who couldn't read station names without stooping.

His solution was to raise the window height into the roof profile, matched by a similar treatment to the doors. In 1949 he showed a

BELOW Car No 10306 photographed in 1971. *(Bob Greenaway collection)*

LEFT Interior mock-up. *(London Transport Museum)*

partial conversion to Lord Latham, who then approved the remodelling of an entire car, completed at the end of the same year. This vehicle then ran in service for 30 years, before being ignominiously scrapped in 1980. It is obvious from the adjacent photograph that interior visibility was dramatically improved (apart from the strange circular windows that were cut into the door pockets); but with Graff-Baker's tragic sudden death on his way to work in 1952, only one of his new ideas was ultimately used; the rest were dropped. No doubt, in those cash-strapped times the estimated increase in cost of £110–£350 per car at 1951 prices to achieve the new glazing had a lot to do with this.

Before Graff-Baker's demise a fully finished half-car mock-up of the planned 1952 Stock design (the programme had slipped a year!) had been completed at Acton Works. This proposed fluorescent lighting, ceiling fans for greater ventilation and a cheaper method of achieving the extra-height windows, using flat glass with the planes meeting at a crisp edge. This shape was continued as a crease line at cantrail level, which in turn was followed through to the edge of the cab. ('Razor-edged' car styling was at the time very fashionable, because aluminium was more readily available than steel and working the material lent itself more readily to this design.) A further proposal specified a bogie design with rubber replacing the steel springs of both the primary and secondary suspension units.

An interesting footnote to Graff-Baker's proposed redesign is that Alec Valentine, who had worked under Frank Pick back in 1928 and was now a board member, had recognised that preserving just the higher door windows would be of considerable benefit to passengers standing in vestibules. This would also substantially reduce the total cost of the 'Sunshine Car', with its revised saloon window glazing; consequently Valentine wanted them kept. However, he was persuaded otherwise by Joe Manser (who at the time had been Graff-Baker's deputy), who said

LEFT Proposed new exterior styling for 1952 Stock. *(London Transport Museum)*

that such a scheme would spoil the side elevation of the new train! What price, then, 'convenience of use' and 'fitness for purpose'?

Ironic indeed, because since the Victoria line trains of the 1960s every tube train type has had them!

1956/59/62 Tube Stocks

After Graff-Baker's sudden death the whole programme was put back, to be reassessed in 1955. But LT's hand was being forced, because new stock was now urgently needed to replace the earliest Standard Stock vehicles, now over 30 years old. It is clear that under this pressure, the simplest, least threatening and most expedient solution was to dust off once more the 18-year-old drawings of '38 Tube Stock. From the outset, however, these new trains would also be left in their natural aluminium finish.

Calculations had shown that a weight saving of only 12% was possible, because the complex nature of the cars' underframe required that it still had to be made from steel. By the middle of 1951 various manufacturers were quoting an oncost of 15% to make the bodies out of aluminium instead of steel, based on an order of 700 cars. A saving of energy costs was proved over a 30-year life, but the benefits were marginal. The reduced maintenance costs that could be achieved by never having to repaint the vehicles over the same timeframe were, strangely, not included!

However, it was agreed that three seven-car prototype trains should be built. Known as '1956 Stock', these went into service on the Piccadilly line one year later. Their successful trial prepared the way for large orders of 1959/62 Stock, the last of which were scrapped in 1998.

In pure design terms they were a step backwards from '38 Tube Stock. Internally, all of the hardware fittings remained identical (except that grab-handles replaced the previous design of bus-type full-length handles on top of the transverse seats), but the original tungsten lighting was replaced by close-coupled fluorescent luminaires that gave a harsh and bland cold light. The original rich, welcoming tones of red and green were replaced by shades of grey, both painted and plastic-laminated, which over the years became

ABOVE The 'new' interior of these trains! *(Bob Greenaway collection)*

BELOW Exterior aluminium style. *(London Transport Museum)*

ABOVE 1960 Tube Stock. *(London Transport Museum)*

BELOW The interior when new. The map/ventilator panels are in the open position. *(London Transport Museum)*

increasingly careworn. Moreover the chosen moquette pattern soon looked grubby, as unlike its historic predecessors it was poor at disguising soiling.

Externally the attractive domed roof of the original with its large glazed destination panel below the cab window was replaced by a blunt, squared-off frontal treatment that incorporated an illuminated roller-blind-operated destination display above the door. This arrangement reverted to what had been previously specified over 33 years ago on the earliest types of Standard Stock!

The 1960 Stock

The train that next appeared on the scene was certainly a step forward in design terms, and although it was never built in great numbers it pioneered some groundbreaking features.

This stock was originally designed to replace all of the old Standard Stock cars operating on the Central line, and 12 driving motor cars were built by Cravens of Sheffield, the first entering service at the end of 1960. Technically it featured two traction motors per bogie connected in permanent series, which removed the need

ABOVE Repainted in '38 Stock livery. *(Bob Greenaway collection)*

for a non-driving motored car. They ran with reconditioned trailer cars of Standard Stock, painted silver to match.

Externally the styling reverted back to a domed roof with ventilation grilles neatly integrated above the destination display. A new 'face' for the stock was created by continuing the side body crease around the front, giving it a slightly raked-back appearance. This subtle change of plane on the car bodies had been a feature on all tube trains from 1923 onwards, its purpose being to tailor the body side profile to clear running tunnels.

Outside and inside, the most innovative feature was the large double-width windows with matching interior casements. These provided a form of double-glazing (albeit not sealed) that allowed the retracting doors to slide into the void created by this form of construction rather than into a localised pocket. Other neat details abounded. Above each window, for instance, ventilation was provided by pulling down a panel which framed the line diagram, whose upper radii matched those of the lower window frame. The ceiling was of peg-board construction (an architectural solution to cut down noise levels), and the fluorescent lighting runs extended unbroken along the car's length.

These trains were intended to be the precursors of a completely new fleet of trains for the Central line, but once again the condition of the oldest generations of Standard Stock did not permit the luxury of a protracted test period, so a repeat order of 1959 Stock was placed instead.

The 1960 Stock then went on to have a chequered history. At the end of 1963 five were converted to trial full Automatic Train Operation on the Hainault to Woodford section. Invaluable experience was thus gained with this cutting-edge technology, which at the time led the world. It enabled the expertise acquired to become a key component of the forthcoming trains for the new Victoria line that were then being specified. A special track-recording train was completed in 1987, consisting of two 1960 Stock motor cars and one later 1973 Stock trailer, to analyse track condition and highlight where repairs or renewal were required. Finally, one three-car unit was painted in 1938 Stock livery in 1990 to commemorate the 125th anniversary of the Ongar line and ran in this condition until the branch was finally closed in September 1994.

Chapter 9

The Victoria line

This was the first new tube railway to be built under Central London since the early 1900s. However, unlike all of these and the other new lines built since 1915, it was designed with a clear objective: to provide improved and more convenient travelling conditions within London's central area and its burgeoning suburbs. This was to be achieved by creating interchanges with many busy stations on different lines, and in particular by linking four of the busiest main-line termini (Victoria, Euston, King's Cross and St Pancras).

It was first mooted as 'Route C' in a package of recommendations put forward by the LPTB directly after the end of the war. The BTC selected it in 1948 as the preferred route of those proposed and gave the London Transport Executive the green light to start promoting it; but the required finance was not on the table. It was clear from the start that the new line could never be self-supporting, but a strong case could now be made for its future role in combating two types of congestion: traffic on the streets and crowding on station platforms and passageways. In 1959 a million pounds was obtained to build twin tunnels along a section of the proposed route between Finsbury Park and Seven Sisters to experiment with new types of tunnel construction and lining.

At this time the Conservative government, in spite of Harold Macmillan's famous slogan 'You've never had it so good', was becoming increasingly concerned about rising unemployment levels, particularly in the North-East. In 1962 some horse trading between the government and prospective contractors resulted in the promise that some of the tunnel segment work would be shopped out to these disadvantaged areas, which is how the Victoria line finally got the green light to proceed in August 1962. This was for the 10.5-mile route from Walthamstow to Victoria; a further 3.5-mile extension under the Thames to Brixton was authorised in 1967.

Preliminary work at Oxford Circus started one month later, and major works got under way early in 1963. The line was opened in stages:

- Walthamstow Central to Highbury & Islington on 1 September 1968.
- Through to Warren Street by 1 December 1968.
- The section from Warren Street to Victoria was officially unveiled on 7 March 1969 by Queen Elizabeth II (the first reigning monarch to travel on the Underground).
- The final extension to Brixton was opened on 23 July 1971, though Pimlico Station would have to wait a further 14 months before it opened because of protracted negotiations with the Crown Estate, major landholders in the area.

Rotary drum digger shields were used to excavate the clay in about one third of the running tunnels, but traditional Greathead shields were employed for the remaining sections, with men working with power-driven hand tools. No mechanical shields could be used on the Brixton extension because of the inherent dangers of tunnelling through the loose, water-bearing sand and gravel beds south of the river.

The trains' operating technology

The Victoria line was designed to incorporate a high level of automation both in train operation and fare collection. The experimental work carried out on the 1960 Stock had born fruit and was written into the specification for new trains ordered from Metro-Cammell (244 cars comprising 122 motors and 122 trailers). For the first time the 'train operator' acted as both driver and guard, first opening and closing the doors at stations and then pressing two start-buttons triggering the automatic control mechanism. Acceleration and braking were then controlled by coded electrical impulse commands that passed through the running rails. A manual override was fitted but would not be required during normal operation, and the driver could be in constant touch with a new line control centre located at the top of a new office block in Cobourg Street, near Euston.

Automatic fare collection (AFC) was achieved through the installation of ticket barriers (or gates) operated by tickets printed with an encoded magnetic strip. Just like their modern successors, the ticket was 'read', allowing the paddles to be opened and entry or exit to be achieved. Virtually all of this dazzling new technology was developed by LT's own engineers, and the automatic trains continued to provide outstanding service to Londoners until they were finally replaced in 2011. Unfortunately the original pioneering AFC equipment was not up to the rigorous demands of daily use and had been taken out by 1972, but the valuable experience gained, coupled with rapid advances in electronic technology, enabled later installations from the mid-1980s onwards to be more than up to the job.

Enter Design Research Unit

This company has its origins back in 1943, when it was proposed by its founding fathers, Misha Black and Milner Gray. Rising to prominence after the war, it played a key role in designing pivotal post-war exhibitions such as 'Britain can make it' in 1946 and the 1951 Festival of Britain. By the early 1960s they were the largest multi-disciplinary design company in Britain, so they were a natural choice to work alongside London Transport's chief architect Kenneth Seymour.

Stations

The budget for these was always going to be tight, and this can still be seen immediately in the type of tiling that was used. The high-quality biscuit and cream tiling of yesteryear, with its 'fit for purpose' bull-nose edged corners where required at high traffic points, was replaced by essentially domestic products in two unremitting shades of grey, with only black and blue square corner details fashioned from lapped tiles with a hard edge. These would soon crack and break. Following much criticism – particularly from the architectural press – that the stations looked bleak and uninviting with their overall 'greyness' and harsh strip lighting, Misha Black is on record as having somewhat defensively stated that the travelling public themselves would provide the necessary colour!

Platform station name roundels were initially made from frosted glass backlit with fluorescent tubes, but these have since nearly all been replaced by enamelled metal versions. To give some relief to 'the rather severe platforms' (LT's own description), attractively coloured tiled niches incorporating solid wooden seats were arranged along their length. Designed by some of the most celebrated graphic designers of their day, they featured for the most part visual references to places and historic events

ABOVE Pioneering automatic fare collection installation on the Victoria line. *(London Transport Museum)*

BELOW Typical platform design at Seven Sisters. *(London Transport Museum)*

LEFT Poster showing 12 of the seating niche tiled designs. *(London Transport Museum)*

above ground. Sometimes where the designer's imagination or source material could not provide an appropriate image, puns were utilised; Brixton, for instance, was depicted as a ton of bricks, and Warren Street as a maze!

The trains

The sleek, uncluttered external styling of the 1967 Stock was a great credit to LT's engineers at Acton and the influence of DRU's Misha Black and James Williams. Fortunately they had developed a good working relationship with Stan Driver, the head of the Rolling Stock Design Office at Acton Works. It chimed perfectly with the world-beating automated drive and control equipment that it contained.

The higher windows set in the doors finally made their appearance (they had continued to be strongly promoted by Alec Valentine, who by now had been knighted and was chairman of London Transport, which is why he finally got his way). The wraparound cab windows looked very stylish and modern and the clean body profile was enhanced by the fact that no drivers' cab side doors were fitted (to prevent the train operator from leaving the cab while the operating equipment was set to automatic). The train whistle was cleverly integrated in one of the front door grab-handles. Double casement windows (from the 1960 Stock) were again specified, and the ventilation grilles were neatly positioned on top of them.

Design Research Unit's terms of reference for their involvement had been loosely drafted, their remit being to produce only concept and general arrangement drawings for the design of the interior. Presumably to save money, they were often excluded from meetings with the manufacturer to discuss implementation details, and this showed in the final design, which was only a very small step forward from 1960 Stock, and in many respects the details were not as well considered. DRU did, however, develop the

LEFT Misha Black and LT engineers compare the new with the old. *(Jim Williams/DRU)*

novel two-tiered armrests that enabled adjacent seated passengers to have their own support, and managed to recess the fluorescent light fittings into the ceiling profile.

The idea to augment these with illuminated advertising panels was not a successful solution; it was pushed through by others, and their design had nothing to do with DRU. Indeed, Misha Black was on record as stating that they would not enhance the interior design of the car and should not be adopted unless the financial benefits were considerable. Whether they were or not is now open to conjecture. These panels differed in size and presentation to the non-illuminated ones and thus produced an uncoordinated appearance, worsened because advertisers tended to supply just one smaller size of advert rather than two. Also, in the fullness of time dead lighting tubes weren't replaced and the framed interiors built up a visible layer of the all-pervading tunnel dust. Passengers could also see an accumulation of the same dirt when standing in front of an opened ventilation panel carrying the line diagrams.

Misha Black had wanted a brand new moquette design to signal the fact that these were brand new trains for a brand new line, but he was overruled; 'we already have too many' he was told. So by default the black, red and grey pattern from the 1961 'A' Stock was used instead.

In spite of its shortcomings the Victoria line was nevertheless a bravura achievement and a great credit to all of London Transport's engineers who worked on it, as well as to Metro-Cammell of Birmingham who delivered trains of exemplary build quality. It had pushed available technology to the absolute limit on a very tight budget. Built at a cost of £91 million, it was carrying passengers at the rate of 82 million per annum just a year after it was fully opened.

Postscript

Today the integrity of the complete Victoria line as built has been compromised, as the design of many of its stations has been tinkered with. Apart from the updating of the continuous name frieze to Henrion, Ludlow and Schmidt standards (see page 142) – a sensible development – walls have been retiled in other colours, and localised ad hoc 'safety initiatives' have been carried out by station managers. At the time of writing only Pimlico remains as originally built and fitted out; even the backlit glass station name roundels have survived.

ABOVE The interior when new. *(Bob Greenaway collection)*

Meanwhile, the decade also saw significant changes in the control of London Transport. In 1962 the British Transport Commission was abolished and London Transport became a board directly responsible to the Ministry of Transport. The Board was required to provide an adequate public transport service for London and to pay its own way. After 1965 this second requirement could not be met, because the government had decided to postpone fare increases that were necessary to meet rising operating costs. Thus started the spiral of 'deficit financing' whereby in order to make up the revenue shortfall LT became, for the first time, partially subsidised by the Treasury. The debate about what represented an acceptable level of subsidy from the taxpayer increased following another change of control in 1969.

From 1 January 1970 the financial and major policy control of London Transport was transferred to the Greater London Council (GLC), which had replaced the previous London County Council in 1965. LT's annual budget was now to be set by the GLC from County Hall, although Central Government would also play a role in part-funding of capital projects by way of 'infrastructure grants'. In this way the completion of the final leg of the Victoria line to Brixton had been achieved by 75% of the costs coming from the Government, with the remaining 25% from the GLC.

WESTMINSTER
Way out →
WESTMINSTER

PART THREE

The third period: the 1970s to date

OPPOSITE Westminster Jubilee line platform with its characteristic cast-iron 'infil' plugs set into the structure. Also visible is the back-lit glass name frieze. *(Mike Ashworth/London Underground)*

Chapter 10

The 1970s

Under the terms of the London Underground's transfer to the GLC the government had agreed to write off LT's accumulated capital debt, which had escalated dramatically as a result of building the Victoria line. It was hoped that this way the organisation could once more be self-supporting, meeting its operating costs from fares income and only requiring financial help from Central Government or the GLC for major capital projects. But this idealistic notion proved impossible to sustain due to the adverse economic climate of the 1970s. Both the Labour and Conservative organisations at County Hall authorised revenue subsidies to LT during the decade, but deciding on the actual level of financial support for Underground fares was soon to prove a contentious issue. This in turn developed into major political and legal squabbles to determine what constituted the 'right' policy for public transport in London in the early 1980s. The 'Fares Fair' policy, introduced by a strongly left-wing political faction within the GLC, triggered off a sequence of events that ultimately led to the greatest overhaul of London's transport operations since 1933, and indirectly to the demise of the GLC.

BELOW Hatton Cross platform design. *(Wikimedia Commons – Sunil 060902)*

The Heathrow Link

The period started promisingly enough with a flagship project in which London would become the first city in the world to have its international airport directly linked to its centre by an Underground railway. This was to be realised by a 3.5-mile extension from Hounslow West, and in July 1970 the £25-million project was approved by the GLC. The Government would not earmark any funds at first, but work started in April 1971 with a 25% grant from the GLC. Over a year later the Government eventually did match County Hall's contribution, but the remaining 50% had to be raised by LT as a capital loan from the money markets.

The extension was built mainly by cut-and-cover to Hatton Cross, a station immediately on the airport's periphery, which was, and remains, a particularly dour and undistinguished design, with a final 1.25 miles of twin-tunnelling under the runways to the heart of the airport. Here an underground station was built close to Terminals 1 and 2, with moving walkways accessing all three of the airport's terminals. The Queen opened the final section from Hatton Cross to Heathrow in December 1977 and by the end of its first year of operation the extension had been used by eight million passengers, confirming that the direct access to many parts of London offered by the Piccadilly line was just what passengers wanted.

Generously-sized island platforms were designed for three stations. Two of these were brand new – Heathrow Central (later named Heathrow Terminals 1, 2 and 3) and Hatton Cross. Two new platforms in the same style were built at Hounslow West, adjacent to and to the north of the existing three platforms there. The dreariness of the Hatton Cross exterior building was relieved within by a sequence of three interlocking 'speedbird'

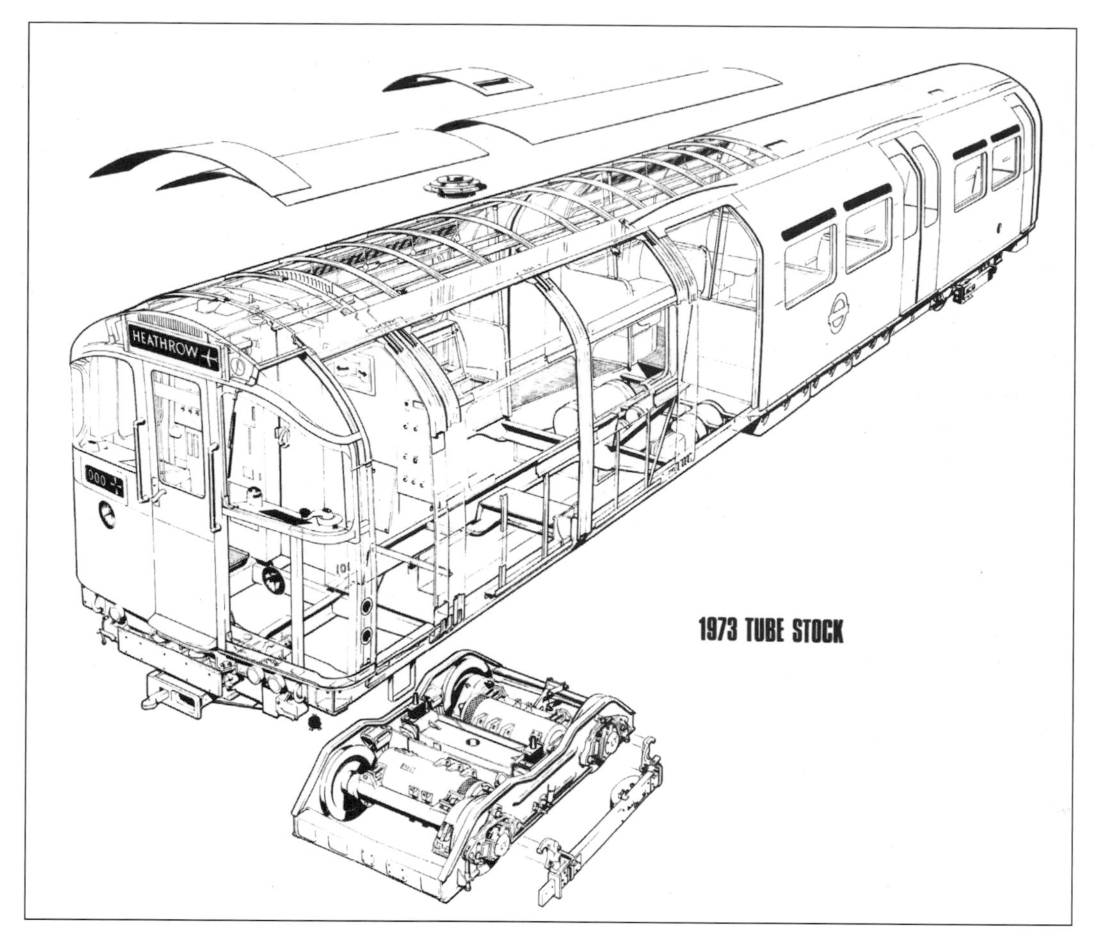

LEFT Stewart Harris' cutaway of 1973 Stock. *(Susan Harris/ Andy Barr)*

motifs set into the orange mosaic tiling of the supporting pillars between the platforms. (Introduced by Imperial Airways on their aircraft in 1938, the 'speedbird' made its final appearance on British Airways Concordes up to 1984, after which it was removed following the introduction of US-designed livery by Landor Associates.) After designing three of the tiled niches on Victoria line platforms, Tom Eckersley was commissioned to produce a similar decorative feature for Heathrow Central based on Concorde tail fins. All of the platforms featured a suspended ceiling that rolled down to a centrally located spine, neatly integrating the continuous station name frieze with the tops of the dividing piers.

New stock for the Piccadilly line

The 1973 Stock trains were a strange blend of excellence and mediocrity in design. In terms of their external styling, they were in their original form almost certainly the best-looking tube trains ever built, with many subtle design niceties to delight the eye.

After the Heathrow extension was authorised, there was initial soul-searching regarding which trains should be used. Although still relatively new, the 1959 Stock trains then running on the line were considered to be unsuitable for the extended service that would generate considerably more traffic. More trains would be required and the luggage of air-travellers would need to be accommodated; therefore new trains were the answer, and the 1959 Stock was in time transferred to the Northern line.

Design work on the new stock began in early 1970 and an order was placed with Metro-Cammell later the same year for a total of 87 trains of six-car length together with one extra three-car unit earmarked for the Aldwych shuttle. Each car was to be 6ft (1.52m) longer than those of earlier types and the door leaves were to be top-hung. On all previous Tube Stock the weight of the doors had been carried in bottom tracks, by means of either flanged

ABOVE One of the trains when new. *(Bob Greenaway collection)*

wheels running on a rail inside a gap within the door, or rollers acting on a bearing surface on the doorstep. Air-operated cab doors were also specified for the first time on a tube train.

Design Research Unit were once again called in to advise on the external appearance, and their industrial designer James Williams, working closely with the Acton design staff, did a first-class job. The expensive curved cab windows of the Victoria line trains were retained but their outline was subtly reworked.

An article in the April 1987 edition of the *Architects' Journal* by the erudite design and architecture commentator Martin Pawley celebrated the style of these trains and extolled the perfection of their detailing. He wrote:

'The advanced engineering of the 1973 Tube Stock, designed specifically for the Piccadilly line Heathrow link and the first made to a life-cycle costed specification, was a breakthrough in the design of underground railways. This combines with the timeless styling of its low profile, clean lines and simple aluminium finish to represent the highest and most exhilarating achievement of functional modern design in the Underground.

'The 1973 Tube Stock has another distinction apart from its advanced engineering. Notwithstanding design parameters as rigorous in their way as those applied to 12 metre yachts and jet airliners, and an operating environment no less punishing; the Piccadilly line train is a tremendous and timeless styling success. The full beauty of the trains can best be seen on the long surface runs out to Heathrow, where they can touch 60mph and their big, automobile style wrap around windscreens, long cars and low tunnel-fitting profile and natural aluminium finish still lend them the electric excitement of a secret prototype – a decade after the line was opened.

'The true merit of its design resides in the tiny arts of artifice that accompany its first class engineering, enabling that which is already simplified to appear even simpler than it is.'

Two of these details especially caught his eye:

'The most subtle example of enhanced simplicity is the "lost" reveal that accommodates the decreased width of the ends of the cars. Each car is tapered over the last metre of its length, where the end doors are situated, in order to accommodate the tight track curves of the line. In plan the taper is achieved by an elegant "lost" rebate that also accommodates the termination of the hollow recesses which form the door retraction recesses.

'Another, even more crucial to the appearance of the whole train, is the dummy ventilation extension above the cars' windows. On the preceding Victoria line ATO (Automatic Train Operation) trains, the ventilator slot above the car windows stops short at the commencement of the door retraction reveal. On 1973 Stock, the black slit of the ventilator on the outside is carried across the retracted door position in the form of a black painted recess in the aluminium outer skin. The result is a visual logic that enhances the engineering and operating logic of the cars themselves. It may pass unnoticed by the million people a day who travel on the Piccadilly line, but once you become aware of it you never fail to appreciate the subtlety of the design of Underground trains again.'

BELOW The 'lost reveal' and extended ventilation slot. *(Bob Greenaway collection)*

But sadly this fine design quality was only skin-deep, because the interior design was no match for it at all. Once again the powers-that-be at Acton kept DRU's designers out of the loop with a 'we know best' attitude. At that time there were no marketing people to create a vision or write a brief for these trains to present to the design consultants, even if they had been allowed to accept it. The opportunity to develop an aspirational modern interior that met the needs of, particularly, airline passengers was therefore lost, and the resulting interior design was totally derivative with no fresh thinking whatsoever.

If the problem of how best to deal with the accommodation of luggage stowage had originally taxed the engineering designers at Acton there was no evidence of it. The rather inept solution was just to specify standbacks by every doorway marginally deeper than usual with a scant amount of space for parked bags and cases. From the beginning this was not a popular solution because passengers found themselves visually and physically separated from their valuable luggage in a crowded train.

The interior details were poorly resolved as well. Ugly ventilation fans (three per car) soon stopped working; and when they were finally decommissioned the housings remained. Sharp-edged castings were to be found where smoothed off shapes would have been more 'user-friendly' and the overall appearance of the interior had an uncoordinated look to it. Moreover the trains rapidly looked scruffy, with the chosen blue/green moquette pattern designed by Marianne Straub soiling badly and soon becoming faded and well worn.

After 20 years of hard use these interiors were, quite frankly, a poor advertisement for London Underground and an unimpressive welcome to arriving visitors from around the globe. The substantially modified and improved post-refurbishment versions (see page 153) were to arrive just in time.

The Jubilee line

Started in 1971, this was another new tube construction project that had a particularly chequered period of gestation. Like the Victoria line, it grew out of initial proposals made in 1949 by the London Plan Working Party who had also recommended a cross-town tube railway to connect with the Baker Street to Stanmore section of the Bakerloo line, the first stage of which was to run through to give much needed respite to its central area between Baker Street and Oxford Circus. By the late 1960s this had grown into detailed proposals called the Fleet line. Parliamentary powers were granted in 1969 for Stage One, which would run in new tunnels to Charing Cross via Bond Street and Green Park. The GLC approved the scheme in July 1970 and pledged to meet 25% of the capital expenditure subject to the Government funding the rest. The following month they did just that, announcing a grant to cover the remaining 75% of the capital cost, and work started near Bond Street only six weeks later.

In 1977 the works were well advanced and tunnelling actually did get within 400yd (365m) of Fleet Street. However, the GLC changed

ABOVE The half-hearted solution for stowing luggage! *(London Underground)*

BELOW The indifferent interior. *(Bob Greenaway collection)*

the name of the line to mark the Queen's Silver Jubilee, though it was impossible to open the line in that year because of a series of delays caused by contractors. It was finally opened by Prince Charles two years later, on 30 April 1979.

The original proposals had shown two further stages of the new line. The second stage was to extend east of Charing Cross to Fenchurch Street via Aldgate, Ludgate Circus and Cannon Street. The third stage would go out to Lewisham via the Surrey Docks and New Cross. However, the money was not forthcoming to realise these plans, so Charing Cross had to remain the line's terminus. A full 20 years would have to elapse before a totally new set of political pressures and commercial opportunities would create an exciting platform to push the Jubilee line eastwards.

More new tube trains for the lines

Originally another fleet of brand new trains was planned for the Jubilee line, and Design Research Unit was once again called in to create design concepts. Illustrated here is an early rendering by pioneer British industrial designer Peter Ashmore. On this, the wraparound cab windows are replaced by subtly curved ones set into a softly profiled front. The notion of a red 'capped' front end and the red doors would reappear 20 years later as part of a standardised corporate livery. The silhouette roundel was a DRU recommendation and was used on some Tube Stock types for a number of years, and this together with the red doors – but no red front – were specified for what became known as 1972 Mark 2 Stock.

BELOW The Peter Ashmore rendering. *(Jim Williams/DRU)*

Because of the now truncated nature of the Jubilee line brand new designs never saw the light of day, which led instead to the birth of what became the 1972 Marks 1 and 2. These were the result of political manoeuvring in high places, including the need to keep Metro-Cammell afloat and the urgent requirement to replace the oldest 1938 Stock cars.

The Northern line was increasingly earning its sobriquet as 'the Misery line', a nickname coined by *The Evening Standard* newspaper. Certainly at the time the name was well deserved, due to the frequent breakdowns of its vintage '38s, the long-term industrial action at Acton Works which inhibited the supply of the vital rotary air compressors for them, and the increasingly decrepit state of many of its stations, coupled with rising crime rates.

In May 1970 Horace Cutler, the then flamboyant, bow-tie wearing chairman of the GLC's Policy and Resources Committee, decided to investigate the problems for himself and took a trip in the rush-hour. He berated the top brass at 55 Broadway to put together a hurried scheme to buy 30 new trains 'just like those on the Victoria line'. These were known as 1972 Mark 1 Stock, and their interiors were exactly like those of the Victoria line trains. With no time for any appropriate redesign they were rushed into service. However, the Northern line was not equipped for automatic train control, and also guards were still carried. The absence of cab doors was soon found to be acutely inconvenient in service use because (unlike on the Victoria line) crew changes would also occur at intermediate stations. This often resulted in the ill-fated drivers not being able to get out of their crush-loaded cars! When they appeared on the Jubilee line 'bodges' were fashioned so that drivers could leave by the central cab door and reach the platform using grab-rails and step plates; however, this was not possible on the Northern line because of shorter platforms!

The purchase of another set of 33 trains was authorised by the GLC in November 1971; these were known as 1972 Mark 2s. This order would enable new trains to be available for the opening of the Jubilee line, and would give the Bakerloo a modern fleet as well. The interiors of

LEFT Baker Street, Jubilee line platform. *(fotoLibra)*

this second batch differed from the first order only in that they featured blue armrests instead of red and the same blue/green moquette pattern as 1973 Stock; this was also fitted on the C69/77 Circle line trains.

Since two-person operation was normal at the time, both batches still had to have the traditional guard's operating control panels fitted at the trailing end of the passenger compartment of driving motor cars. Additionally, since it was expected that one day the Mark 2 trains would eventually run on a one-person-operation automatic Jubilee line, door controls were also fitted in the cab. In the event both the Jubilee and Bakerloo line trains were finally converted to OPO form in 1988, but they never went automatic.

In the early 1990s some of the Northern and Jubilee line Mark 1s and Mark 2s were transferred to the Bakerloo to augment their fleet, and received their same major interior design refurbishment. At the time of writing (2013) these are still performing well and their appearance and cleanliness in service is a great credit to the operators of that line.

One-person-operation trains

In the ever deteriorating financial climate that perpetually prevailed, one-person-operation trains had been a goal fervently desired by LT's management for a number of years owing to the large staff cost savings that would accrue from the elimination of guards from trains. All the new trains that were delivered after 1968 were designed to be suitable for OPO, but it was not until the early 1980s that agreement was finally reached with the unions over its introduction. A lengthy process then ensued over a number of years to fit all station platforms with mirrors and TV monitors. Their positioning had to ensure that drivers always had a clear field of vision along the entire platform length before starting their train.

Interior station design

Obviously stung by criticism levelled at their 'lavatorial wall' grey-on-grey Victoria line platforms, this time around Design Research Unit – working with chief architect Sydney Hardy – really pushed the boat out, with very bright colour schemes. Bond Street was tiled in deep cobalt blue and Charing Cross in lime green, which was Sir Misha Black's recommendation (he had been knighted in 1972 and died in 1977). Baker Street and Green Park were tiled in flame red.

As can be seen from the photograph of Baker Street, the platform design emphasised the inherent 'roundness' of its construction, with the only vertical elements being the vitreous enamelled grey housings for the station name roundels and the grey entrance and yellow

CLOCKWISE FROM TOP LEFT The four seating niche designs: Baker Street, Sherlock Holmes *(Sunil 060902 Wikimedia commons)*; **Bond Street wrapped present** *(Elf at Wikimedia commons)*; **Charing Cross, a photographic image taken from Nelson's Column** *(London Underground)* **and Green Park leaf** *(Wikimedia commons – Oxyman)*; **and . The Jubilee line platforms were closed after the new extension opened and have been recently seen in several films, such as *Skyfall* (2012) where they deputised as 'Temple Station'.**

exit portals. A useful safety feature was the highlighting of these exits by bands in the same yellow colour that wrapped around the tunnel barrel, effectively contrasting with the more numerous grey ones. A strip of this same yellow carried the 'Way Out' messages above the station-name frieze, a neat example of coordinated design. (Such a relationship was not used on Victoria line platforms; the continuous frieze on these had its 'Way Out' messages in white out of black with only a tenuous relationship to the illuminated 'Way Out' signs.)

As would be expected, design details were a development of and more simple in execution than those seen on the Victoria line platforms. The grey station-name frieze, with the name in black, ran continuously from end to end; it provided not only indirect lighting for the tiled walls and the decorative seat niches but was also designed to hide cabling and trunking. The decorative elements were:

- Baker Street – scenes from various Sherlock Holmes tales illustrated by Robert Jacques.
- Bond Street – ribbon-wrapped packages in bright green and cobalt blue.
- Green Park – dark brown leaves on the flame red background designed by June Fraser, then of DRU.
- Charing Cross – eight different photographic images of Lord Nelson taken by David Gentleman.

Charing Cross station amalgamated with the old Trafalgar Square and Strand stations. Its Northern line platforms (formerly Strand) were superbly modernised with end-to-end black-on-white murals by David Gentleman depicting the medieval Eleanor Cross on laminated panels. It was the first of the many platform refurbishments that followed and it still is arguably one of the very best. It was unveiled at the same time as the official opening of the

new Jubilee line by Prince Charles at the end of April 1979. He admired both of the new platform designs, but was kept way from the decrepit Bakerloo line platforms. An equally dramatic and successful modernisation of these soon commenced using decorations by June Fraser and Richard Dragun (also of DRU). They created 350ft-long laminated murals illustrating paintings held by the National and National Portrait galleries. However, because of budget constraints this work was not finished until 1983.

The work at Charing Cross came with a brand new ticket hall with a brightly illuminated suspended ceiling. The feature colours of cobalt blue and lime green were used on columns and wall surfaces.

New Surface Stock designs

Circle line 'C' Stock

In May 1968 an order was placed with Metro-Cammell for a total of 212 cars, making up 35 six-car trains. These became known as C69 Stock and were all in service by the end of 1971 on the Circle and Hammersmith & City lines. Their delivery enabled the remaining 'CO/CP' to be transferred to the District line, allowing all of the remaining 'Q' Stock to be scrapped. A further order was later made to replace the oldest of the 'CO/CP' trains to operate the Wimbledon–Edgware Road branch of the District, and these were delivered in the middle of 1977.

In visual design terms, they were in every way a poor design and at best looked dull and workmanlike, without any distinctiveness in their styling. Designed as a central area 'people-mover', each car had four sets of doors, allowing for rapid entering and exiting over the short distances between stations. They did contain some useful operational features now employed for the first time on any Underground train. These were: independently controlled train operator's cab doors with their own air supply, and a 'selective close' facility that enabled all but one set of doors to be closed at terminal stations. These aspects were later incorporated on the new Piccadilly line trains as well. Eight transverse seats were fitted between draught screens that effectively 'walled off' the interior into separate cubicles with poor through-car visibility, and transverse bulkheads carried illuminated advertising panels. Seats were trimmed with Marianne Straub's blue/green moquette pattern, originally designed for Victoria line trains but never used on them; at least it harmonised well with the light blue laminated surfaces trimmed with satin anodised extruded aluminium sections.

The interior ambience suffered badly over time. The 'Aerowark' material covering the heater panels peeled away, exposing adhesive stains, and the illuminated advertisement panels were often left unlit and, as on the late 1960s Tube Stocks, showed a build-up of tunnel dust. During the 1980s they were badly hit by the graffiti menace inside and out, which made them appear hostile and threatening. The

ABOVE LEFT David Gentleman's artwork. *(Bob Greenaway collection)*

ABOVE The related Bakerloo line platforms. *(Bob Greenaway collection)*

ABOVE 'C' Stock external appearance. *(London Transport Museum)*

BELOW Original interior. *(London Transport Museum)*

compartmentalised interiors promoted crime, particularly when passengers were travelling late at night. In a later chapter we will see how, as with other stock types, a major refurbishment in the late 1980s could not have come too soon.

An LT Design Survey Report of June 1971 reveals how Design Research Unit, as the appointed design consultants, had been deliberately excluded from all aspects of the design process. The report said:

'The exterior of these trains is retrogressive in its design as it re-establishes the low browed look in spite of the success of the high forehead of the Victoria line driving cabs. At present design initiation is largely the result of collaboration and proposals between London Transport and Metro-Cammell engineers and design proposals for the exterior form and interior treatment are submitted to the Design Panel when these have virtually reached finality and only minor modifications can then be proposed. We were not consulted at all except being invited to offer comments on the colour of the interior, which was in any case a "fait accompli"!'

This situation illustrated how much had changed since the days of the Underground's international reputation for excellence in vehicle design; that rich legacy appeared to be virtually in tatters. It demonstrated that the Design Panel, under Misha Black's chairmanship, was at the time ineffectual, and also that not a single highly placed LT executive was insisting on and ensuring that exemplary design was important.

New trains for the District line

That Design Survey Report must have caused an internal rumpus, because DRU were this time involved in the design of the trains. Known as 'D78' Stock, they would replace the entire historic family of types still operating on the District, and the first complete train of the type entered service on 28 January 1980.

The 'high forehead' was back, and the cars had a commendably clean body profile. The cab design was simple yet elegant and well-proportioned, and carried the same partially painted red 'apron' as used on '73 Stock. Three-and-a-half foot wide single-leaf doors were fitted with passenger-operated door-open buttons. (These had been tried on and off since 1936 but had always ended up being taken out of use for reasons of intermittent reliability;

even with the modern control technology that prevailed by 1980, they would suffer the same fate this time round too.)

There was an international influence to the design because DRU were at the same time consultants to Metro-Cammell for the Hong Kong MTRC railway, and their styling proposal for the trains shows similar sheer body sides. Since each car was 60ft long (18m), some 8ft longer than their 'R' Stock predecessors, they had to be made narrower to negotiate every curve on the line. The bodies were built as wide as possible at floor height, but had to be slightly tapered to roof level.

At the time DRU had working for them Jürgen Greubel, a young German industrial designer who had worked for Dieter Rams, the celebrated designer of many iconic Braun electrical products. He developed and produced a scale model which defined the design approach.

The interiors were also a great step forward. They were very warm, bright, cheerful and welcoming – all very positive attributes. The colour scheme was a very attractive solution using flecked oatmeal laminates, accented by the then very fashionable colours of orange and brown. The moquette design (by Sir Misha himself) harmonised nicely and featured four colours – mustard yellow, orange, brown and black – in a 'stacked bricks' format.

Draught screens were commendably low, which gave an open and airy aspect and good visibility from end to end of the car, which had become an important safety feature. Finally all of the design elements, such as the extruded ventilation grilles, lighting troughs and framing of line diagrams and advertisements, were well integrated and nicely detailed.

And so the decade came to an end. The redecorated platforms at Charing Cross would herald many more such schemes during the 1980s, some excellent, others certainly less so. But by the end of the 1970s even casual users of the Underground system would increasingly notice disquieting examples of a lack of management control, which in the 1980s would have disastrous consequences. These ranged from sloppy housekeeping to poor regulation of retail premises. Even Piccadilly station, once the jewel of the system, had been allowed to degenerate into a safe haven for dossers and drug users.

ABOVE 'D' Stock train when new. *(Bob Greenaway collection)*

BELOW The original attractive, welcoming interior. *(Bob Greenaway collection)*

Chapter 11

The 1980s

During 1980 the Conservatives were in control at County Hall, and in a desperate attempt to stem the ever-increasing operating losses on the Underground as well as the buses, fares were increased twice in one year – by nearly 20% in February/March, and six months later by an additional 13%. Not surprisingly, this move hastened the annual declining ridership even further, from the 720 million journeys made in 1948 to 498 million journeys in 1982.

The Underground soon found itself a political punch-bag, as the cost of travel in London became a key GLC election issue in 1981, and when a Labour-controlled GLC was swept into power its manifesto included a promise to cut fares by 25%. Instructed by their new masters, LT therefore introduced the 'Fares Fair' scheme, which also introduced for the first time a system of zonal fares throughout the Underground and bus networks, with an average fare reduction of up to 32%; back to where they started, then. Passenger usage on the trains and buses immediately rose from 5.5 to 6 million a day, and a welcome by-product was a downturn in car usage and traffic congestion in and around London.

All of these benefits had to be funded by an additional £125 million a year from ratepayers, which increased the level of London's public transport subsidy from 29% to 56%. The validity of this additional rate demand by the GLC was challenged in the courts by the London Borough of Bromley, which did not have the luxury of any Underground services! They won their case and 'Fares Fair' came to an end in March 1982, with the Law Lords pronouncing that London Transport had to plan, so far as was possible, to break even financially. To enable this to happen, fares were then increased by 100%, with the inevitable result that ridership plummeted back to 5 million a day! This was obviously a step too far, so a compromise position was reached in May 1983, when fares were cut by 25%.

The long overdue zonal fare structure was retained which was the launch-pad for fresh marketing initiatives as typified by the 'Travelcard', an idea borrowed from the Paris Metro. This was available for one day, seven days, one month or a year and gave travellers the freedom of both the Underground and bus services within one or a combination of specified zones. This was successful from the start, for there was a 40% increase in Underground journeys from 1982 to 1985.

In 1984 the Conservative Government of the day removed London Transport from the control of the GLC (which itself only survived a further two years) and a new corporation called London Regional Transport was formed in June of that year, directly responsible to the Secretary of State for Transport. Its mandate was to improve London's transport services whilst halving the level of revenue support from tax and rate payers to £95 million by 1988. This would be achieved by more efficient and cost-effective operating methods, such as a new ticketing system, one-man-operation of trains and devolving maintenance to the respective depots.

London Underground Limited was formed in 1985 under the companies act as a subsidiary company of LRT but still in public ownership.

Rolling stock developments

The most dramatic changes to train design occurred during the 1980s, which began inauspiciously with a very humdrum design known as 1983 stock. Fifteen trains of this type were ordered for the Jubilee line in June 1982 from Metro-Cammell, comprised of 30 three-car units all of the double-cabbed type. They were judged to be smaller versions of 'D' Stock, with which they shared many features. Design Research Unit were by this time out of the picture and the new trains were designed

completely in-house at Acton Works. No industrial design consultancy was involved, and quite frankly it showed!

The cab had missile-proof flat glass panels set into a totally flat front with particularly sharp corner radii. No subtle shaping or finessing here! The same single-leaf doors as 'D' Stock were also incorporated. The moquette design was likewise inherited, but its feature colours were turned into a very badly judged interior scheme. The attractive oatmeal colour was replaced by bilious mustard that permeated the whole interior, relieved only by orange on the door surrounds. The design of the ceiling fan housings was overly complicated and they looked remarkably ugly.

One plus point in the design was a continuous row of luminaires running from end to end which gave exceptional even lighting. Unfortunately providing adequate sealing against tunnel dust was once again a problem, and on a repeat order of 16½ trains in late 1986 to augment the service, lighting reverted to good old bare tubes.

The 1983 stock had a very short service life, as scrapping of the first variants started in earnest a mere 15 years later. Generally no one was sorry to see them go. There were many operational difficulties with the single-leaf doors, and there were always mechanical problems associated with the bogies, together with ongoing electrical problems that exasperated line engineers were constantly 'fixing'.

By the early 1980s the original Design Panel had become the LT Design Committee, with considerably more power and influence. Because of the generally poor design of the 1983 Stock, they insisted that henceforth design consultants had to be fully involved right from the start. The Department of Architecture now had 'design' added to its responsibilities with the appointment of a design manager.

Industrial design consultants are introduced

When the signalling on the Central line needed replacing by the late 1980s, it was also decided to bring forward the scrapping of its 1962 Stock, replacing it with brand new trains and thus harmonising both major projects. The

ABOVE 1983 Stock showing the effects of graffiti attack. *(LURS collection)*

BELOW The distinctly unattractive interior. *(Bob Greenaway collection)*

ABOVE The new (for London Underground) all-welded extruded aluminium construction. *(London Underground)*

signalling was to be replaced with an updated version of the Victoria line's ATO system and the line's traction supply was to be boosted.

A review was made of suitable design companies who could work closely with the rolling stock engineers at Acton, and in October 1982 DCA Design Consultants of Warwick were chosen to submit design proposals. Factors in their favour were the ability to build full-size mock-ups near their premises and the fact that David Carter, the founder of DCA, also employed mechanical and electronic engineers and so could provide an all-embracing service.

It was a time when there were great changes abroad in the technical development and construction of rolling stock. In the late 1970s British Rail had done much pioneering work on their prototype APT tilting train using larger load-bearing extruded aluminium body sections welded together, the technology having being developed by the Swiss company Alusuisse. This manufacturing method was of great interest to the Underground's engineers because a complete car body could now become load-bearing, as well as being stiffer and lighter, whereas previously all bodies had been coach-built and mounted on separate underframes, either steel or aluminium. There was also an opportunity to explore technical innovations such as air suspension, motors on all axles, pressurised ventilation for passenger saloon areas and electronic 'thyristor' control of the operating current. On some of the cars steered bogies and motor-mounted disc brakes would be tried.

The three prototype trains

It was decided that three different prototype trains would be built to test out many of the options available. DCA therefore produced three alternative designs – the red 'A' train, the blue 'B' train and the green 'C' train. In 1984 orders were placed to build three four-car trains at a value of £11.7 million. Metro-Cammell built the red and the green trains and BREL (British Rail Engineering Ltd) the blue one. There was one set of control equipment for each two-car unit (eight motors). Regenerative, rheostatic and friction braking were available, or blends of two.

Doors had to be externally hung since the extruded aluminium construction no longer allowed for 'pocketed' internally sliding doors, because large 'flying buttresses' around the

BELOW The three prototypes. *(London Underground)*

RIGHT Interior number one – the red train. *(Bob Greenaway collection)*

CENTRE Interior number two – the blue train. *(Bob Greenaway collection)*

BOTTOM Interior number three – the green train. One driving motor car survives in the LTM Reserve Collection at Acton. *(Bob Greenaway collection)*

door apertures had to be welded in to maintain torsional stiffness and rigidity.

The external styling and the interior design of all three variants is illustrated, and market research findings which identified the best features from each of them were written into a final specification for what would become the Central line's 1992 Stock.

Stations

In general terms the focus on the replacement of rolling stock during the 1970s had left little money available to improve and update station environments. In fact hardly any modernisation of the predominantly Victorian and Edwardian central stations had taken place for 40 years.

The new Charing Cross Northern and Bakerloo platforms were rightfully judged to be a great success, so in September 1981 the GLC authorised a £60-million (at 1981 prices) rolling programme to revamp a large number of stations. A sizeable proportion of this budget was earmarked for some of the busiest central area stations such as Bond Street, Embankment, Holborn, Piccadilly Circus, Tottenham Court Road and Waterloo.

Unfortunately the amount of money available was eventually cut back from what was originally on offer, so instead of total refurbishments this dictated that for the most part only the platforms themselves would receive a cosmetic update. This created a contrast between careworn old and sparkling new, which gave a somewhat confusing message to passengers! The main difference was now that the former uniform approach that had survived from Edwardian days was replaced by a bespoke scheme inspired by

ABOVE The restored original platforms from 1863; the replica globe light fittings have still to be fitted. *(Bob Greenaway collection)*

features from the local vicinity. For most the designs very successfully brightened up the platforms and made them more cheerful and inviting places in which to wait for trains. As had been already realised at Charing Cross, Baker Street, Tottenham Court Road, Piccadilly Circus, Paddington and Embankment stations were able to go considerably further as they received all-embracing schemes.

Baker Street

For many years the historic Metropolitan and Circle line platforms that dated right back to 1863 had been covered over with murky panelling. When stripped away it was found that the original brickwork had been preserved in reasonable condition, so a fine job was done in restoring these features. The ventilation shafts were white tiled and fitted with indirect lighting to replicate the original daylight that would have flooded in. All the new works at Baker Street were completed by 1983.

Elsewhere around the station complex, new tiling bearing the head of Sherlock Holmes was used, and these same images were cleverly used to form large silhouettes either side the of platform roundels above seating bays on the Bakerloo line platforms. A key part of each scheme was the large trunking housings in line colour designed to house extraneous cabling.

Having supplied such generous casings, the in-house architects were often mortified to discover that during the preceding night the communication engineers had screwed their new cables *above* these ducts! The traditional lack of communication between departments, particularly when using sub-contractors, has ensured that this practice is by no means dead!

BELOW Revamped Baker Street platform. *(Copyright Mike Quinn, Creative commons Licence)*

Tottenham Court Road

This major rework was a fully blown design solution that then, as now, polarised opinions; it is loved or hated in equal measure!

The sculptor Eduardo Paolozzi had been commissioned in 1979 to design a series of mosaic murals covering the whole station that would reflect the music stores, hi-fi retailing, entertainment and nightlife in the area. Visually complex and brightly coloured in the extreme, the designs were plastered over platform walls as well as almost to half ceiling height. Passageways, the tops of escalator shafts and concourse areas (with brightly polished aluminium ceilings) all received the same treatment – literally no stone was left unturned!

The particularly flamboyant Central line platforms were opened for business in 1983, but the slightly more subdued Northern line ones opened two years later.

Piccadilly Circus

If anything could be labelled bright and vibrant, the new platforms, network of tunnels and lower concourse areas of Piccadilly Circus were certainly that. Refurbished between 1984 and 1986, below escalator level high-quality glossy cream tiling (with wide grouting lines) was used everywhere, extravagantly highlighted by four feature colours of rust-red, green and the line colours of blue and brown. All of this was particularly brightly and harshly illuminated, and its brassy appearance was accentuated by polished handrails in that material. Subtle it was not!

The abysmal state of the classic Holden sub-surface concourse had been finally recognised and a thorough £19.5 million renovation of the ticket hall was at last carried out. It successfully

ABOVE 'Love it or hate it'! *(London Transport Museum)*

LEFT A new look for Piccadilly Circus platforms. *(Mark Ovenden)*

LEFT Superb restoration of Holden's 1927 masterpiece. *(Mark Ovenden)*

RIGHT **Brunel's tunnelling drawings. When this picture was taken, the housings for the cables had still to be fitted.** *(London Underground)*

incorporated the new UTS ticket machines (of which more latter), telephone points and an LT information kiosk and was officially reopened by the Lord Mayor of Westminster in June 1989.

Paddington

The refurbishment of ticket hall, passages, escalators and Bakerloo line platforms featured decorative tiling that used fragments taken from an original patent drawing for tunnelling machines by Sir Marc Brunel for the Great Western Railway. This was discovered by the artist employed, David Hamilton. This theme was very successful because it was subtle and never intrusive.

BELOW **Robyn Denny's artwork.** *(London Transport Museum)*

Embankment

Platforms and passageways in this large station serving three lines were completely re-clad in formed gloss white vitreous panelling. (This technique was also being increasingly seen on European metros such as Berlin and Vienna.) The artist Robyn Denny was commissioned to produce artworks to relieve and enliven their chilly blandness and this resulted in a series of coloured 'streamers' that were intended to give a 'festive' flavour. The station was completed in 1988.

Some other notable station redesigns

Holborn

Also completed in 1988 was the complete reworking of both the Central and Piccadilly line platforms at Holborn. Here the pre-war biscuit-cream tiles were also covered by large vitreous enamel panelling, in this case displaying monochromatic images by the artist Allan Drummond of artefacts held in the British Museum, and also medieval images by Christopher Tipping showing minstrels and dancers. The black, white and grey tones made a rich contrast to the station roundels together

LEFT The reworked Central line platform. The Northern line platforms received a similar treatment. *(London Transport Museum)*

with the trunking housings and seating in either line red or blue.

Bond Street

This scheme, completed in 1983, incorporated a 'wrapping paper' motif (to suggest the retail activity topside). In places this graphic treatment rolled boldly up and over the tunnel barrel to the baseline of the trackside wall. The wall tiling also continued down horizontally to form the flooring; truly a fully integrated design solution. Although it could be said that this design might have merited a place in the 'future design heritage', it only lasted 24 years, being completely stripped out in 2007 and replaced by white 'lavatorial' tiling, thus returning to its original 1903 appearance. The reason cited was cheaper repairs and easier sourcing of replacement tiling.

LEFT Bond Street platforms of 1983; sadly no more. *(Roger Hall collection)*

Heathrow Terminal Four

When a fourth terminal was planned at Heathrow, it had been decided to build a further tube extension to serve it. The line was opened on 1 April 1986 by the Prince and Princess of Wales. Its clean-lined 'international' style single platform giving immediate access to the ticket hall represented a new design approach. It was simple yet immaculately detailed, the only decoration being the large incised numeral '4' set into the cream-coloured, reconstructed marble walls.

BELOW Heathrow Terminal Four. Note the numeral '4' above the bay of 'tractor' seats as originally fitted. *(Bob Greenaway collection)*

ABOVE A typical signing before HLS' involvement. *(Chris Ludlow)*

The new Underground signing system

All of this work triggered an internal squabble between the in-house Architectural Services Group and the then Department of Signals and Electrical Engineering, who by default had become responsible for the design and installation of all new signing, both illuminated and non-illuminated. Much of the signing on the station modernisation works was not to the agreed 'standards', they said. When challenged to produce these, they presented a scruffy A4 piece of paper with some notional rules and examples of layouts in Johnston capital letters! These were based on historical precedent, the varied inheritance of a chequered past which was then freely interpreted.

As the 1980s progressed it had become increasingly obvious that there was a large gulf between the perception of the Underground's traditional commitment to design excellence and the actual reality 'on the ground'. This was now particularly apparent in the way that stations were signed. The clarity of the high-quality signing from the 1930s Piccadilly line and the 1940s Central line extensions was still to be seen relatively untouched, and the Victoria line had maintained the tradition of ordered and considered grouping of signs. However, the busy interchanges in Central London told a totally different story.

No proper guidelines existed for the design control of signs, and as a result visual anarchy prevailed. New signs were unceremoniously slapped over or placed adjacent to old ones, and the inconsistency of an endless variety of graphic solutions could be seen all around the system. This layered accretion of haphazardly designed, executed and installed signs prevented a controlled sequence of information being presented to passengers and didn't convey the impression of a modern, efficient train service.

Henrion, Ludlow and Schmidt, a leading international corporate identity consultancy, were commissioned in 1983 by the Architectural Services Group to prepare an intensive report on the current status of Underground signing. Unsurprisingly this identified, with a fresh pair of eyes, countless defects and irregularities. Chris Ludlow took the lead on the project and he made some fundamental recommendations:

- Signing must be part of a system that is optimally functional.
- A planning and positioning guide must be created which will set criteria and procedures to be followed.
- Positioning of signs is currently very haphazard. They should aim to be appropriate and consistent in position and application. This means that passengers should not have to search for them in a busy visual environment.

HLS were subsequently commissioned to produce designs for a new signing system that was installed in Victoria Station, the busiest on the Underground system, during the middle of 1987. Market research found that it performed admirably and it was ratified for system-wide implementation. Fine surviving signs of historical design merit would be retained as long as they still performed operationally.

The key attribute conveyed was 'consistency in every manifestation' and nowhere is this more apparent than in the design of every 'Way Out' graphic device, so important for the safe and speedy evacuation of stations. Illuminated signs are always yellow out of a black field set within a standardised white panel (essential for reading in a smoke filled environment). The same treatment appears in non-illuminated form on predominantly station-name friezes. Previously 'Way Out' signs were presented in a variety of ways, sometimes illuminated and otherwise in a variety of different colours and backgrounds.

HLS' recommendation to set up a dedicated internal sign planning and management unit was also acted upon, and this team exists to this day.

The cornerstone of the system is a standardised methodical and wholly consistent approach to all the design elements. A decision was made to replace the Old Johnston typeface, invariably still used in its upper case form, with the New Johnston family of faces. These had been designed by Banks and Miles in 1980 for use on LT's promotional material and general printed typography because they gave greater versatility than simply using the old display face. Upper and lower case letters were now to be used on all directional signing because years of research had proved them to be more legible than capitals. This is because words are mainly identified and read by their distinctive outlines, rather than letter by letter.

Capital letters were, however, retained in the Underground roundel, and for station names appearing in the bar of the roundel, fascias and frieze runs. A modular approach was now to be taken to overall sizes, layouts, and constructional details. Finally, line colours would now be in a consistent proportion and position on each sign to confirm orientation from ticket hall through to platform.

The signing system has given a remarkable degree of added value to the Underground. Its friendly approachable character is quite different to the somewhat aloof style of the old signs. Research has shown that the ambience of a station appears fresher and cleaner when fitted with new signing, and no better example of its excellence can be seen than in the new signs in the Paris Metro, which has been heavily influenced by Ludlow's work.

Arising from this successful work, Henrion, Ludlow and Schmidt were also appointed as the corporate identity consultants for the Underground by the then marketing and development director and marketing manager, to develop a suite of standards and guidelines that developed the same

ABOVE Signing to the new standards. *(Chris Ludlow)*

BELOW New-style directional signs in action. *(Author's collection)*

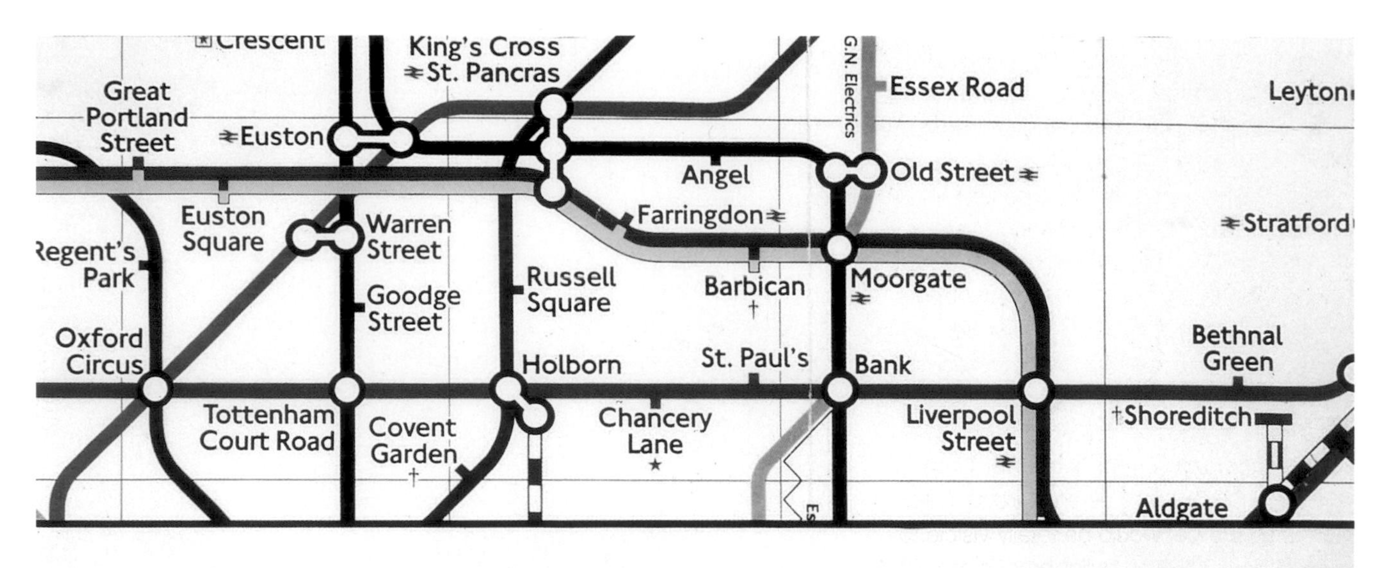

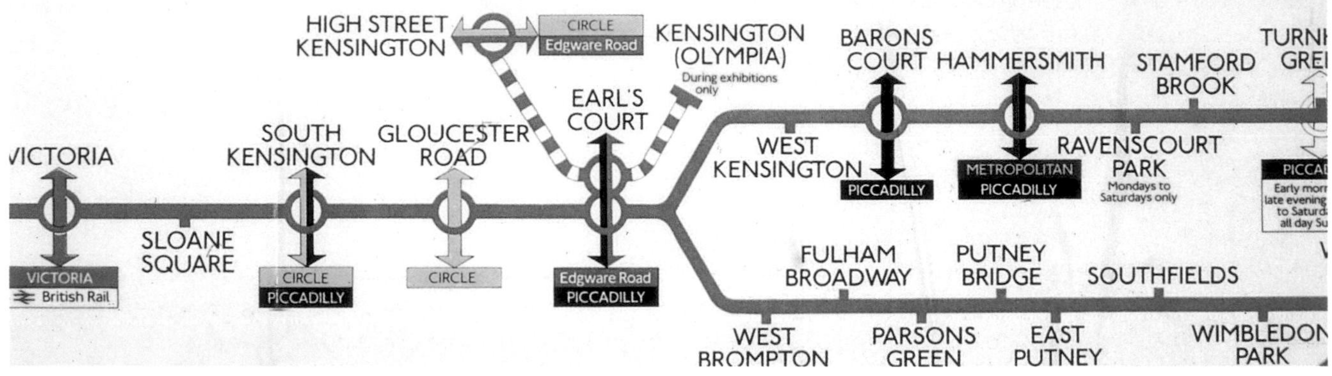

ABOVE The previous graphic inconsistencies between the map and the in-train line diagrams. *(Chris Ludlow)*

BELOW Harmony restored! *(Chris Ludlow)*

messages of clarity and consistency across every visual manifestation of LT's rail business. This work included inter-alia maps and in-train line diagram standards, publicity standards, train livery standards and emergency vehicles standards.

Passenger security initiatives

Crime became an increasing problem during the decade, particularly on stations at the southern end of the Northern line. A trial installation of passenger 'Help Points', giving information and assistance (wired through to the British Transport Police), was carried out at these stations and a number of others. These prototypes featured off-the-peg 'World War Two U-boat' style hardware mounted on a rectangular background. However, it soon became apparent that these did not stand out sufficiently on a crowded platform, as they blended too readily with maps and posters.

Colin Cheetham Design Partnership were brought in to solve the problem, and the result was the now familiar 'pill' used around the system (and also at a large number of Network Rail stations). Immediately successful, the new design's distinctive round form stood out in a crowd; furthermore, market research discovered that in the hours of darkness on open air platforms, passengers would gather round them, reassured by their presence.

A further revision of security arrangements had been to develop 'Focal Points' where a staff member, in sight of monitors from a group of CCTV cameras and directly linked to the BT Police, would be totally visible to passengers – behind bullet-proof glass. DCA Design Consultants designed the booth, and it was to be positioned near to platforms or within a ticket hall along with passenger seating, with 'Help Points' and dot matrix 'Next Train' describers close at hand. Prototype installations were made but they never went system-wide. One of the design aims was to make staff members more visible to passengers; traditionally they had always hidden behind ticket office windows, obscured even further by a proliferation of stuck-on notices. In recent years, however, fully glazed ticket office windows have become the norm, looking equally at home on the most modern stations and listed 1930s examples (*eg* Acton Town).

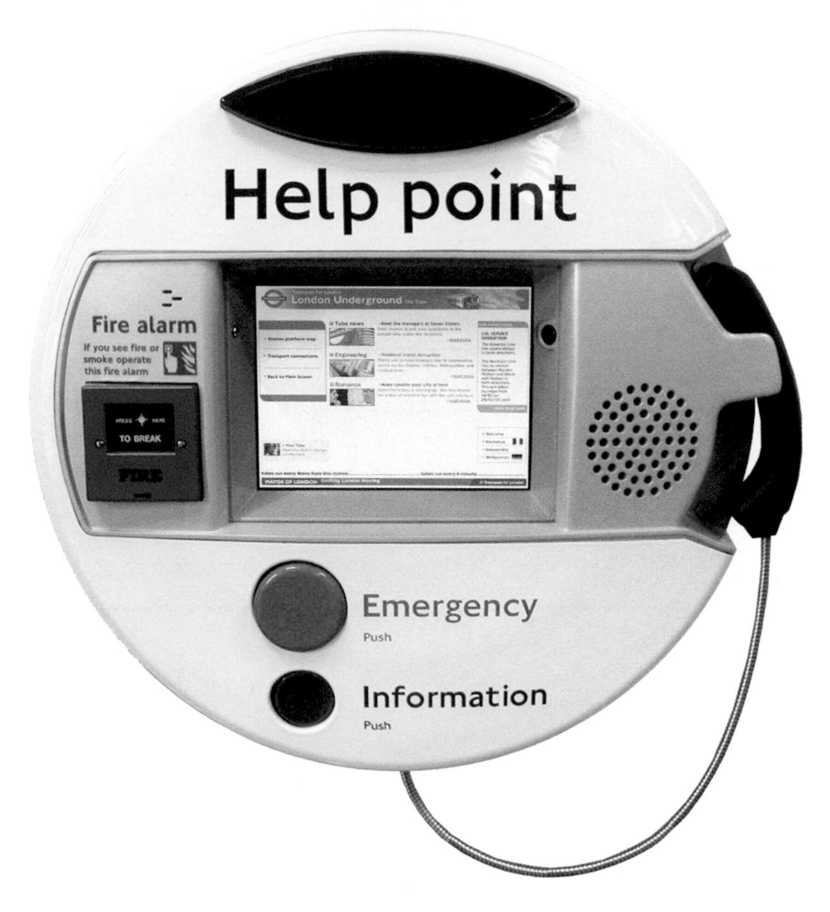

ABOVE Proposed enhancement of the product by John Elson Design. Trialled at Gloucester Road, it features a telephone handset and a graphic information display screen. *(John Elson Design Company)*

BELOW LEFT The trial installation. *(DCA Design International)*

BELOW The original 'pill' installed at Ladbroke Grove, still there today. *(Author's collection)*

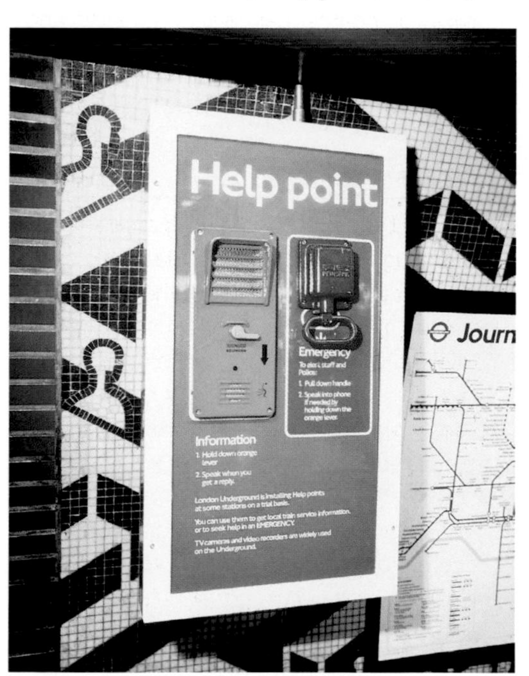

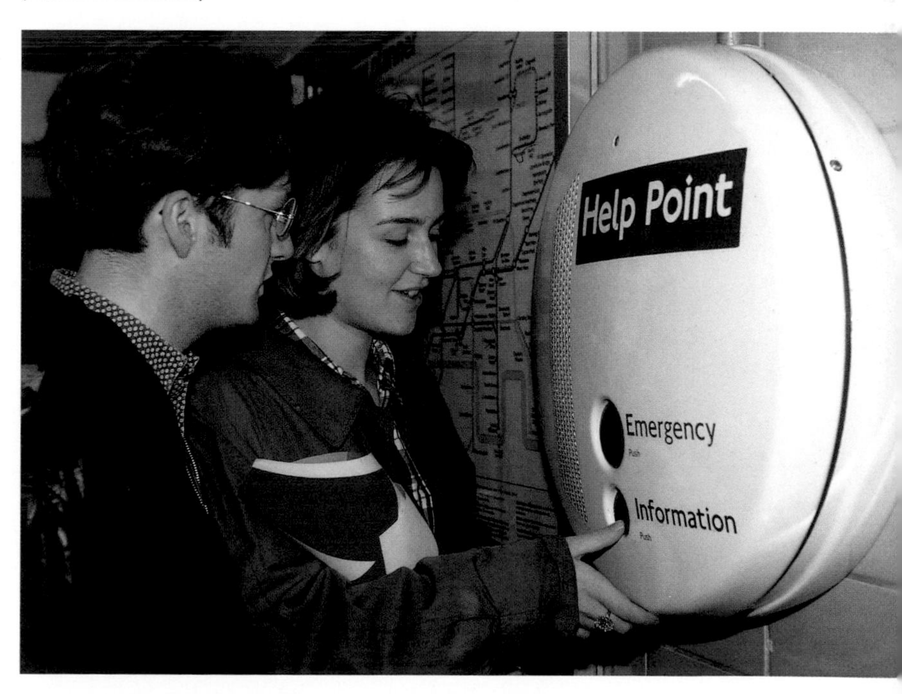

ABOVE The dot matrix train describer. *(DCA Design International)*

The first electronic indicators fitted on the Northern line were in crudely fashioned rectangular boxes, and were notoriously unreliable in that the information given was often wildly inaccurate! Nevertheless, by the mid-1980s further development of the electronic software had cured this problem, and later versions were housed in an elegant sculptural housing designed by DCA. Their convenience and safety aspects were a welcome bonus for passengers waiting on platforms. Subsequent versions have reverted to the rectangular box format, but these include a digital clock as well.

Dot matrix train describers

The first electronic 'Next Train' indicator was installed at St James's Park in 1981, and for the first time showed a countdown of the number of minutes that passengers had to wait for their train. Its antecedents were still to be seen around the network, some dating from 1910 and others from the 1930s and 1940s, but these only showed the order of arrival for the various routes. (Two fully restored Edwardian ones can still be seen in everyday use on the Earl's Court sub-surface platforms.)

The Underground Ticketing System (UTS)

During the 1980s armed robberies directed at ticket office staff who held their cash in free-standing booths (called passimeters) had been increasing alarmingly, as had the amount of revenue lost through fraudulent travel. The fatal shooting of a staff member at the south end of the Northern line was the catalyst for immediate action to resolve the crisis.

From 1987 onwards, the UTS initiative involved the rebuilding of all of the ticket halls to provide 'secure suites' that featured new wall-mounted ticket machines with bullet-resistant windows behind which staff could handle money transactions and cash-up in safety.

This programme initially created considerable hostility with interest groups such as English

RIGHT Wall-mounted ticket machines. *(London Underground)*

ABOVE LEFT The classic electro-mechanical machines dating originally from 1937. *(London Transport Museum)*

ABOVE The Mk 1 gates. *(London Underground)*

Heritage and 'The Thirties Society', who felt that many fine historic ticket halls were being needlessly destroyed in order to accommodate a standardised layout for the secure rooms. In the event the solutions were for the most part very sound, with careful replication of the original tiled finishes wherever possible. Indeed, some of the results were excellent solutions and are a great credit due to the unsung in-house architects who achieved this thankless programme within very tight budget and time constraints.

The new ticket machines finally replaced the vintage free-standing, grey, automatic ticket machines, with their angled glass fronts, that had been a characteristic feature of ticket halls since before the war. Complementing them was a new design of stainless steel-clad automatic gates (which enjoyed the benefits of much-improved operating technology than had been available for the Victoria line in the 1960s), with associated gate-line hardware. Both the wall-mounted machines and the first series of gates (initially fitted only to the Zone 1 central area stations) were designed and built by Westinghouse-Cubic Limited, with DCA employed as industrial design consultants by London Underground.

The 'Few Fare' machines were simple to understand and operate, but the first generation of 'Multi Fare' ones, using the known technology available at the time, presented passengers with a myriad display of buttons, each one marked with a station name in small print. Predictably these caused delays and confusion for passengers when buying their tickets. When the Jubilee line extension team was later established, they wanted something better for their multi-

LEFT A sea of station names on the multi-fare machine! *(DCA Design International)*

RIGHT The touch-screen ticket machine designed by Andrew Parker. *(Cubic Corporation)*

fare ticket machines, so WCL responded with touchscreen activation in various languages, a far simpler and more easily understood solution. When the budgets became available, virtually the same design was retrofitted across the rest of the network and is now the standard solution seen everywhere. Similarly, a touchscreen version of the 'Few Fare' machine has been introduced in the last few years.

Another priority was the installation of gates on every single station, and for this the elegant Mark 2 'slimline' gates with a visually related barrier system were used. These were designed for WCL (now known as Cubic Corporation) by Andrew Parker, their then industrial design consultant.

Six years of operating experience together with the findings of the Mott-MacDonald safety report on the Underground Ticketing System were blended with updated technological, design and ergonomic opportunities to create a dramatically improved product. Both versions are still to be seen across the network. Much careful thought was given to a redesign, and key improvements were:

- The stanchion width was slimmed down to enable seven gates to be positioned within the same space as six of the original units.
- The requirements for stored value ticketing were anticipated within the design.
- A less intimidating, more approachable image.
- Incorporation of the Underground's requirements for reverse breakthrough, electrical rather than pneumatic paddle control, and improved information displays.
- 'Stepforms' were avoided wherever possible to inhibit gate vaulting. The glazed central infill panel psychologically deterred vaulting as well as giving a visual lightness to the design, making it less intrusive in a ticket hall.
- Paddles were reduced to only two per side, as opposed to four on the original design. The paddle shape was altered to reduce the possibility of catching clothing and to ease the release of trapped items.
- Finally, when open the paddles now parked flush with the sides of the stanchion walls, to maximise the usable aisle width.

RIGHT The Mk 2 slimline gate. *(Cubic Corporation)*

The King's Cross fire

The tragic and disastrous fire which commenced under a wooden escalator in the late evening of 18 November 1987, killing 31 people, revealed the managerial depths to which the company had plummeted since the halcyon pre-war days of Pick and Ashfield. The poor housekeeping that had been visible over many years around the system but simply ignored had turned to bite them.

Understandably this was a true watershed for the company, and quite rightly things would never be the same again. There was a litany of disasters that exposed the senior management as flabby and distant, with in particular no real understanding of the importance, management and implementation of proper fire detection measures. After the accident the management system was reorganised and modernised. Denis Tunnicliffe, the new managing director recruited from British Airways, soon discovered that in many areas there was a management desert; the entire system seemed to operate on a daily basis through sheer happenstance. In its place he would soon create a coherent management structure based in part on practices borrowed from his previous employer. To break the organisation down into manageable chunks, each line was now organised as an individual business unit, and 'centurions' were created, station and train crew managers each responsible for 100 staff. No longer were promotions achieved only through long service, irrespective of capability.

The report on the fire by Desmond Fennel QC identified some incredible facts. Unbelievably, from 1939 to November 1987 there had been 123 reported incidents of fires starting below escalators due to smokers' materials setting light to dirt, grease and human hair. In every case these fires had extinguished themselves after a few minutes and thus they were never reported; this had been considered the norm until that fateful night at King's Cross.

On the evening of the fire the ticket hall was seriously undermanned because of sloppy supervision of meal-break times. It was found that the ticket office staff had no training in fire-fighting or station evacuation procedures, regarding their role as being solely confined to ticket office sales. They belonged to a different union to the other front-line station staff.

Money from central Government was soon found to address many of the inadequacies found in the aftermath, the first priority being the installation of state-of-the-art fire detection and control apparatus, together with up-to-date communications equipment and exhaustive training in its use. A massive programme was also undertaken to remove all materials that no longer met the strictest fire safety standards, both in trains and in stations, and this activated a major rolling stock programme that radically improved the trains – effectively making new trains from old.

This initiative will be described later, but first an example of a product that was developed after King's Cross to meet the extremely rigorous fire safety standards.

ABOVE Typical scene before the King's Cross design initiatives. *(London Underground)*

New ticket collector's booth

The new booth's predecessors, manufactured from hardwood with melamine infill panels, no longer met the enhanced fire safety criteria introduced after 1987. The accompanying picture also shows how disorganised and ad hoc the display of passenger information had become! DCA designed an elegant new booth manufactured from extruded aluminium and fabricated sheet metal assemblies mounted on a cast aluminium base. It was crisp and modern in style, expressing the fact that it belonged within a modern railway, and it harmonised very well with all the other gate-line ancillaries. Later on, following the gating of the entire system, these booths became known as Gate Line Attendant Points, in which role they are still to be seen all around the network.

BELOW A brand new image. *(London Underground)*

Chapter 12

The 1990s

As was mentioned above, a major rolling stock refurbishment programme had started almost immediately following the King's Cross fire, but most of the work was carried out during the 1990s.

DCA's work on the 1986 prototypes had shown the great improvements that external design consultancies could accomplish on train interior design; consequently there would be no going back to the engineer-led solutions of the past.

First in line for improvement were the Victoria line trains from 1967. Clearly the replacement of their fibreglass mouldings and melamine-clad surfaces (which gave off cyanide fumes when burning at very high temperatures) provided a great opportunity to simultaneously introduce engineering changes required to radically improve their interiors. Jones-Garrard and Tilney Shane, then working together as part of the Transport Design Consortium, were commissioned to do the work.

The final interior was everything that its predecessor was not, certainly in later years. It had many positive attributes for passengers, being warm, inviting and colourful. Key elements of the redesign were:

- The curved draught screens formed an ellipse with the new ceiling profile, on to which light was cast by new glass luminaires. A warm ivory powder-coated finish was chosen for the interior panelwork, including the doors, with a 30% gloss level to reduce reflections.
- The ivory tint was accented by a bright blue colour, closely related to the line's blue livery as used on grab-poles, handrails and armrests. Grab-poles were mounted clear of the floor to aid cleaning, and the armrests were firmly anchored into new seat frames to resist vandals snapping them off by walking on them, a common problem with the original version.
- A distinctive bold moquette seating pattern was designed, with ivory and blue in its composition.
- The time-honoured ribbed maple wood flooring (a staple component of Underground trains since the beginning of the 20th century) was replaced by a newly developed composite rubberised material finished in two blue tones, the darker colour being used to delineate the door vestibule areas.
- The original car ends were not effectively lit, making these areas look dim and gloomy. Round porthole lights were added to effectively illuminate them, and market research findings showed that passengers responded positively to this detail. They appreciated the fact that it was designed for their well-being and not there just for operational or engineering needs.

Finally the trains were painted in a corporate

BELOW The new 1967 Stock interior. *(London Underground)*

colour scheme using graffiti-resistant paint, because by now this anti-social practice had reached epidemic proportions – hundreds of aluminium-bodied trains had been scarred for life, as the aerosol paint had eaten into their porous structure. There was a highly positive response to the first prototype train that went into service in 1989. Passengers felt that it displayed renewed operator pride in the rolling stock, and that they themselves were also being valued; in fact it was considered good value for money for their taxpayers' pounds!

The 1972 Bakerloo line trains

Since these trains were virtually identical it followed that their redesign followed a similar path but with a strong line identity. Thus a terra-cotta colour replaced the blue with a colour related moquette design. The colours harmonised really well and were probably the best modern re-incarnation of the warm and welcoming character of 1938 Stock. At the time of writing they can still be seen operating on the Bakerloo line but the moquette has been changed and the armrests removed.

The 1969/77 Circle line trains

This was an even more dramatic reconstruction of these trains, as can be seen from the 'before' and 'after' pictures here. Cre'Active Design, an offshoot group of ex-DCA designers, were chosen as the consultants. They replaced the intimidating walled-off interior 'cells' with a particularly bright interior, with fully glazed draught screens giving exceptional through-car visibility, enhancing active and passive security. This feature was augmented by the installation of car-end windows, just visible in the photograph! The same ivory shade was chosen as the interior shell colour, highlighted by yellow grab-poles and handrails with two grey colour tones used on the new composite flooring, doors, armrests and draught-screen castings. Seats were now all longitudinal, with a reinforced anti-vandal steel mesh built into the moquette.

At the time of writing these trains are still in service but are gradually being replaced by a brand new design of Surface Stock (see later Chapter).

LEFT The Bakerloo line variant. *(London Underground)*

BELOW The hostile-looking original interior with graffiti and peeling panelling. *(Author's collection)*

BOTTOM The redesigned interior – bright, open, welcoming; everything the old vehicles were not. *(Bob Greenaway collection)*

ABOVE The final redesigned interior. *(DCA Design International)*

BELOW Finally, passengers with luggage are well provided for! *(TDI/Warwick Design)*

The 1960/62 Metropolitan line trains

In 1988/89 DCA were appointed by Metro-Cammell to work with London Underground on a heavily revised four-car unit to explore the potential of new refurbishment ideas. As with 'C' Stock, the profile of the heavy bulkheads was greatly reduced to give a brighter, more open ceiling, and new draught screens incorporating horizontal grab handles and continuous luggage racks were fitted, as well as new hopper windows. However, in 1991 available funds were very thinly spread, and refurbishment plans stalled. In fact at one stage changes were only going to be on a like-for-like basis, with no funds left for any redesign; but meticulous project management, blended with DCA's continuing involvement, meant that ways were found to achieve most of the redesign work at the same prices. ABB at Derby won the contract to transform the fleet and the first train was delivered during 1993. Much of the original concept was kept, but the deep bulkheads and individual hat racks were retained (suitably repolished), as were the all-important car-end windows.

The redesign was very successful, and the colourful, welcoming and cheerful interior was light years ahead of the old one. It served these trains well right up to their departure in 2012.

The 1973 Stock Piccadilly line trains

The extensive refurbishment programme finally gave these trains the interiors that they had always been denied. Foreign visitors to Heathrow no longer had to endure the 'shock treatment' of entering scruffy and unkempt trains; instead they were met with a stylish and colourful environment that continued the quality ambience of air travel right into central London. This work was carried out by Warwick Design Consultants, yet another DCA offshoot. The design grew out of a detailed mock-up that they had previously produced for a refurbishment of 1983 Tube Stock for the Jubilee line (in the event, brand new trains were ordered to cover that line's extension requirements).

The project was let to RFS Industries Ltd, who unfortunately went into receivership halfway through the programme, and Bombardier Transportation inherited the contract in 1994. As

ABOVE The complete interior. *(Bob Greenaway collection)*

a result of this change the integrity of the design was soon being compromised, and it was found necessary to create a new management team representing all the parties involved, and a new fully detailed mock-up was agreed and signed-off by all parties.

Car end windows were also fitted, and extra deep stand-back areas finally dealt satisfactorily with luggage. When not required, this space also formed a large multipurpose area to accommodate wheelchairs. In addition, a generously proportioned moquette perch seat that aligned nicely with the seat tops could be used. The ventilation grilles were given a pleasingly soft sculptural form, as were the catenary posts that clamped the horizontal handrails. The redesign also showcased many other well-resolved details that focussed on passenger and operator requirements, such as:

- Six dot matrix train destination displays in each car.
- Lift-up seating bays for easier maintenance.

BELOW Easy and convenient maintenance of equipment. *(Bob Greenaway collection)*

ABOVE The folding detrainment steps. *(LURS collection)*

RIGHT The stylish 'masked' window design. *(Author's collection*

- Folding steps (designed by Jones-Garrard) built into the central cab door for the safe and convenient emergency detrainment of passengers. These replaced the time-honoured wooden ladders traditionally kept in the cab for this purpose.
- Skilful integration of CCTV surveillance cameras.

The painted corporate livery suited the vehicles particularly well, and their smooth appearance was enhanced by a satin silver linear treatment on the inside of the windows, shrouding the multipurpose areas.

The District line trains

This refurbishment completed the set. Partly because the fire safety performance of the original interior was still very good, there was not the same pressure for change as on the other stock types. It consequently took four years for an original design concept by Jones-Garrard from 1996 to emerge as a prototype vehicle, but at the time many of its proposed features could not be afforded and the design was consequently 'de-scoped', preserving just some of them.

Eventually the trains' mid-life refurbishment was carried out under the ill-fated Public-Private Partnership by the company Metronet (who later went into receivership), and the first two units entered service in the summer of 2005. Car end windows were provided and, of course, the maple flooring was replaced with composite rubber.

The previous strap hangers were replaced by horizontal handrails with a gently undulating profile (inherited from previous concepts). These and the unchanged vertical handrails were refinished in a lighter version of the line's green colour. New features were added for the convenience of passengers – a multipurpose area with tip-up seats was provided for wheelchair users, and dot matrix next station describers were fitted laterally throughout the car. This information was also very usefully installed behind car windows adjacent to the doors. Draught screens with their exposed fixings remained unaltered, as did the ceilings and ventilation grilles. CCTV surveillance cameras were also discretely housed.

ABOVE The 'de-scoped' interior; nonetheless still a very successful redesign. *(Author's collection)*

Finally the train operator received an air-conditioned cab. In addition to repainting the bodies in corporate livery an additional measure was added by the application of window film to combat so-called 'Dutch graffiti', *ie* scratching the glass.

New 1992 Stock Central line trains

As was mentioned earlier in this book, the best features from the three 1986 prototype trains were blended into a new design specification that was to become the 1992 Stock train.

A contract was awarded to BREL, Derby Works (British Rail Engineering Limited, which later became ABB Transportation, then Adtranz and finally Bombardier Transportation) to manufacture 85 six-car trains. The large panoramic sealed windows from the green 'C' train were incorporated, while the bays of six longitudinal seats with their continuous 'up and over' grab-pole/handrail came from the red 'A' train.

The middle pair of the six seats in the outer bays was set back 4in to allow greater standing capacity around the grab-pole positioned there; this was made possible by the available floor space freed up by the bogie travel underneath. The trade-off for this was that handrails had to be terminated above the two centre seats so that the heads of seated passenger would clear them on rising. Closed armrests (better called seat separators) were designed with 'frangible' connections that would allow them to break away cleanly if a crush-loaded train was to be involved in a serious collision. Unfortunately vandals were quick to discover this characteristic, and the armrests were soon destroyed by lateral kicking; a suitable more robust replacement has yet to be found. They are, however, still intact on the five sets of four-car trains of the same type used on the Waterloo & City line. Clearly these must cater for a better class of passenger!

ABOVE A 1992 Stock train when new. The 'up and over' windows from the 1947 'Sunshine Car' make their appearance once again, albeit in a totally different format. *(London Underground)*

In recent times such trains have been updated with line-related aqua-coloured grab-poles and handrails – a very pleasing colour and a lot more restful for jaded commuters than the original red.

DCA built a full-size fully finished mock-up to define and agree the final design's appearance, inside and out; this was displayed at the London Transport Museum in Covent Garden in December 1990 to celebrate 100 years of tube trains beneath London.

The externally hung doors were made as wide as possible to create a clear double-door aperture of 1,664mm (832mm on single doors) to optimise boarding and alighting at the busiest stations. The front cabs and all the car ends are capped by large super-formed aluminium sub-assemblies, which

RIGHT The staggered seating bay either side of the grab-pole. *(DCA Design International)*

LEFT The 'Addison' cab design. *(London Underground)*

replaced the large phenolic mouldings used on the 1986 prototypes.

Another Underground first was the fact that the train operator's cab received the same care and attention to detail as the passenger environment. Addison Design Consultants had at the time an accomplished ergonomic unit and they were retained to input proposals for seating comfort and posture, as well as the shaping and positioning of controls and instruments.

The trains featured both automatic train operation (ATO) and automatic train protection (ATP), which effectively allow the trains to drive themselves. The ATO is responsible for operating the train whilst the ATP detects electronic codes in the track and feeds them to the cab, displaying the target speed limits. The propulsion for the trains was manufactured by a consortium of ABB and Brush Traction and was one of the first examples of microprocessor-controlled traction using a fibre-optic network to connect the different control units. The DC traction motors are controlled via GTO (gate turn-off) thyristors.

There were many technical teething troubles with this stock, which is not surprising given the degree of innovation it demonstrated in virtually every area.

From 2011, after almost 20 years of continuous service, these trains were updated both technically and cosmetically. The cab front was subtly reworked to eliminate water ingress and to replace a complex set of parts with a much more simple design. The car windows have also had to be substantially modified, with large surround frames applied whereas they had previously been flush. This was again to prevent water leaks. The new 'Barman' seat pattern has also recently been introduced.

BELOW The 'Barman' moquette, designed by textile designers Wallace Sewell. *(Author's collection)*

ABOVE **Example of the art deco detailing.** *(Mark Ovenden)*

Station design in the 1990s

Inevitably, the huge Jubilee line extension project with its awe-inspiring station designs overshadowed other station projects of the period, but these must also be acknowledged as significant.

Bank/Monument station

A new Docklands Light Railway terminus at Bank station was the catalyst for a major rebuild encompassing the linking together of five tube lines and ten platforms to create a huge complex. Designed in-house, the common visual thread that linked them was the use of the same light grey smoked tiles. These were enhanced at platform level by the City of London coat of arms in moulded relief, which flanked the roundels. In order to give some subtle colour relief to the all-encompassing light grey, an art deco theme was chosen, with the treatment of columns, service doors and lighting units all inspired by that era. The complex was completed in 1996.

Angel station

The first station dated from 1901 and for many years had suffered from increasing congestion and overcrowding; this was particularly alarming for those passengers who gathered during rush hour on its original island platform. Moreover, its old lifts were well past their 'sell by' date, and the whole site looked shabby and neglected.

Consequently the station was rebuilt at a cost of £70 million, opening in September 1992. YRM were the architects. It was integrated within a brand new office complex in Islington High Street, just around the corner from the original station site in City Road. The notorious island platform was filled-in to create an extra-wide southbound platform, and a new section of tunnel was excavated to provide a matching northbound one. Because of the depth from ticket hall to platform, YRM were the architects, and, because of both the great depth from ticket hall to platforms and the need to 'break up' the long access corridors en route, they developed a concept of entering a series of 'rooms', the largest of these being a mid-level concourse accessed by three flights of escalators. These are the third-longest escalators in Western Europe (after those found at Västra Skogen and Kamppi stations on the Stockholm and Helsinki metros), with a vertical rise of 90ft (27m) and a length of 197ft (60m). From this intermediate point a shorter length of three escalators leads down to the lower concourse at platform level. The mid-level concourse was 'safeguarded' in 2007 as an interchange point for the proposed Crossrail Two line between King's Cross/St Pancras and Essex Road. Under its previous name, the Chelsea–Hackney line, this had been a potential new route since the 1970s.

The exuberate decoration of previous stations was replaced by a series of simple neutral tones and finishes in white and coppery brown, all fashioned from high-quality materials of vitreous enamel and marble.

LEFT Angel station platform. *(London Transport Museum)*

Externally the station was signposted by the first application of a new three-dimensional 'cotton-reel' roundel (designed by Derek Hodgson Associates), which has since become a common fitting on prestigious locations such as the new Jubilee line stations, the new entrances at Green Park and elsewhere.

In the future Angel is planned to be an interchange on the proposed Crossrail 2 line, 'son' of the Chelsea–Hackney line that had featured as a theoretical new route in various forms since the 1970s.

Hammersmith

This brand new station with a major bus interchange was built in conjunction with a major island-site retail redevelopment. Minale Tattersfield were the appointed design consultants on the project, which was completed in 1993. It featured glazed canopies over the Piccadilly and District line

LEFT First application on site of the 'cotton-reel' roundel. *(Author's collection)*

RIGHT Artist's impression of Hammersmith ticket hall. The tiled wall is patterned to suggest a reflection of Hammersmith Bridge and is topped by rescued elements of the original station's ceramic name frieze. *(Minale Tattersfield)*

RIGHT The platforms by the same artist. *(Minale Tattersfield)*

platforms, and the two original coloured visuals give an excellent representation of the ticket hall and platforms.

Gloucester Road

Extensive refurbishment with some rebuilding of this original historic station, dating from 1868, was completed in 1993, also in conjunction with a commercial development.

LEFT Restored platforms at Gloucester Road – proof of how well the new signing worked in an historical environment. *(Bob Greenaway collection)*

Chapter 13

The Jubilee line extension

As we have seen, the Jubilee line between Baker Street and Charing Cross was intended to be just the first phase of a line that would extend through to New Cross and Lewisham, and a short extension eastwards almost as far as Aldwych was actually constructed before being abandoned for cost reasons.

Plans to extend the line were resuscitated in the late 1980s, prompted by the massive redevelopment of Canary Wharf in the former Docklands area. The overall scheme by the developers, Olympia & York (then the biggest property company in the world), was a very attractive one to politicians because it far outstripped any other plans to revitalise the then wasteland area. As a result of the deregulation of the City thousands of new jobs would be created in what was to be the largest commercial development in Europe – a financial services district of a million square metres built around One Canada Square (then the tallest building in the UK), housing 50,000 jobs. The Reichmann brothers, owners of O&Y, certainly thought big!

Servicing this huge development could only be achieved with appropriately fast and reliable transport links with excellent interchanges to other transport modes. This, then, was the catalyst for the Jubilee extension – in fact its main reason for being.

London Transport waited for the recommendations of the East London Railway Study before committing to the final route. The study's conclusions promoted an extension of the Jubilee line from Westminster to Stratford via Canary Wharf and Canning Town. This option was decided upon at an estimated cost of £2.1 billion, to which Olympia and York would make a £400 million contribution. In the end the bill would come to £3.5 billion, partly due to huge cost overruns during construction, and O&Y's final contribution would be less than 5% of their original figure.

The line was routed through to the North Greenwich peninsula because British Gas plc had made a contribution of £25 million so that the railway would serve a proposed housing development on land that they owned; and initially it was touch and go whether new stations would also be built at Southwark and Bermondsey, which, as had always happened in the past, might bring much needed regeneration to the surrounding area.

Construction officially commenced in December 1993 and was scheduled to be completed in 53 months. An imported technology developed between 1957 and 1965, called the New Austrian Tunnelling Method

BELOW Building the tunnels with the NATM system. *(London Transport Museum)*

(NATM), was employed, in which a thin layer of concrete is 'shot' around the just-bored tunnel face, and then strengthened by a flexible combination of rock bolts, wire mesh and steel ribs rather than a thicker concrete lining. Since the turn of the 21st century NATM has been used for soft ground excavations and making tunnels in porous sediment. However, a major setback occurred in October 1994 when tunnelling for the Heathrow Airport Express Link from Paddington collapsed in spectacular fashion, opening up a large crater between two runways that caused structural damage to buildings and car parks; this had used the same NATM method. Tunnelling on the Jubilee line extension was consequently stopped for six months pending a report on the new technology by the Health and Safety Executive.

It was also planned to use the technically very innovative 'moving block signalling system' on the extension, which would enable the throughput of 36 trains per hour, with the headway managed by computerised control rather than using traditional fixed signals. But Westinghouse, the designers and contractors, finally admitted that they could not get the software to work. The 'moving block' system therefore had to be abandoned, and was replaced by the long-established 'fixed block system', causing further delays.

The line was eventually opened in phased stages during 1999. Stage one, from Stratford to North Greenwich, opened on 14 May; the connection to Waterloo was opened on 24 September; the connection with the existing Jubilee line was opened on 20 November; and the final stage to Westminster was opened on 22 December, just in time to transport VIPs from Westminster to North Greenwich for the official opening of the Millennium Dome.

Station design

London Underground had never before had such a truly remarkable set of stations built for it, the best of them being quite literally awe-inspiring. When Sir Keith Bright, chairman of London Transport, got his marching orders following the King's Cross disaster, he was replaced by Sir Wilfred Newton, newly arrived from the Hong Kong Metro, which was at the time the newest and arguably the best in the world. He brought with him one Roland Paoletti, an architect who had worked with him in Hong Kong as the creative design supremo for 11 new stations. In spite of the huge project overspend, the architectural legacy that he masterminded for London remains a stunning one, with its extravagant spaces and monumental scale. From the outset the stations would be designed to be 'future-proof', so that they could readily respond to ever-increasing passenger growth levels over the next 15 to 20 years. Much had been learnt from the King's Cross fire (where overcrowding and a lack of spacious exits had contributed to the scale of the disaster), as well as other new metros around the world.

The stations at Waterloo, Southwark and London Bridge had to fit with existing Network Rail and Underground infrastructure, but new stations at Bermondsey, Canada Water, Canary Wharf and North Greenwich were effectively clean pieces of paper on brownfield sites. These would be built within massive concrete boxes, and would draw as much natural light as possible down through the stations to enhance their spaciousness.

Paoletti was keen from the outset that there would be a close interaction between the architecture and the engineering, and decreed that the structures should therefore 'speak' through their architectural forms. There would be no applied decorative elements whatsoever; indeed, an instruction went to the appointed architects to leave civil work exposed wherever possible. The emphasis throughout the stations is therefore on revealing the naked engineering aspect of the constructional details. Some of these have been artfully styled, such as the fluid forms of the ceiling and supporting columns of Canary Wharf; others reveal a more brutal aspect to their detailing.

A different architectural practice was chosen to design each station, thereby stamping an individual character on each, and part of their commissioning process was that they had to demonstrate a thorough understanding of engineering principles.

Key features were to be step-free access all the way from entrances to trains, dual exits at both ends of the stations for safety evacuation,

LEFT Platform edge doors fitted at Westminster. *(Adrian Pingstone, Wikimedia Commons)*

smoke ventilation systems, fireproof lifts and at least three escalators per station, a total of 118 in all – which increased the number operating on the entire network by 40%. Typical finishes and materials used throughout are glass, brushed aluminium and stainless steel, terrazzo, moulded concrete and grey- or blue-enamelled cast iron infill 'plugs' on platforms and elsewhere.

Sir Wilfred had secured ring-fencing of the whole project from LT involvement, and it consequently had its own management, methodology and money. All appropriate existing London Underground standards had to be applied – with, of course, design for fire safety being uppermost – and the now firmly established HLS signing standard was a given. A neat development of the latter was to reinvent the continuous station-name friezes on platforms in backlit satin-etched glass.

Also new to London were the platform edge doors by Westinghouse, whose opening and closing harmonised with the train doors. This was a new feature borrowed from the Singapore Metro that improved airflow and prevented suicides. They were installed in all of the subterranean stations where the long, straight platforms (untypical of so many on the rest of the network) made for their convenient installation. In the author's opinion, the most notable of the stations are:

Westminster

This was totally transformed by architect Michael Hopkins, who had been an obvious choice because his design for Portcullis House adjacent to the Houses of Parliament was already being built above the station. This was a notoriously difficult site to work with, because the District and Circle lines operated as normal

BELOW Westminster. *(Author's collection)*

LEFT Southwark ticket hall – a nod to Arnos Grove. *(Author's collection)*

CENTRE Southwark – the intermediate concourse area. *(London Transport Museum)*

while the largest hole in Europe was being dug around them.

In an immense 40m (130ft) deep cavern-like environment reminiscent of Fritz Lang's silent film *Metropolis*, the concrete and steel structures with their supports are revealed in all their naked glory while passengers create a rhythmic motion between them as they intermingle through the spaces on their escalators.

Southwark

Designed by Sir Richard MacCormac, this is arguably the most stylish and inventive station of them all. It initially eschews the concept of light entering downwards because the ticket hall – a nod to Holden's Arnos Grove – is totally illuminated by uplighters apart from a small area around the central spine. However, the top of the escalators is irradiated by natural light drawn from a circular drum fashioned from glazed blocks. Passengers then proceed down to an intermediate concourse that opens out to a 16m (52ft) high void which is liberally illuminated by daylight from above. On one side is a huge concrete wall clad in cast masonry blocks, through whose portals the escalators enter. Its counterpoint is a gently curving 40m (131ft) long wall made from 660 pieces of specially cut blue glass, supported on a steel framework by means of cast steel 'spiders'.

Bermondsey

This station was designed by architect Ian Richie, who was one of the first to be chosen by Paoletti because he was, in his words, very 'structurally minded' and would collaborate wholeheartedly with structural engineers. Now that a decade has passed since these

LEFT Bermondsey – bare concrete walls but plenty of natural light over the escalators. *(Author's collection)*

LEFT Canada Water – the glazed drum. *(Author's collection)*

stations were opened with much ballyhoo in the architectural press, a more objective view can be taken of them. Unlike Southwark, the overall aesthetic of Bermondsey station appears somewhat harsh and charmless, although its location in a previously neglected part of London has clearly spurred major property development.

The street elevation in Jamaica Road seems to lack any appropriate 'railwayness' in its appearance, being more akin to a glazed rectangular supermarket building that happens to have an Underground roundel affixed to it! Inside, the exposed concrete walls of the concourse areas, escalator walls and metal accents have a stark functionality to them, as they are stripped right back to the barest essentials of finishes. The escalators, however, receive an abundance of natural light drawn from a fully glazed roof, and Richie has created a rich contrast to the monotonal finishes by using moulded blue glass for the platform seating.

Canada Water

This station designed by Herron Associates, with a particularly fine adjoined bus station by Eva Jiricna Architects, continues the theme of impressive scale throughout, and although some of the areas might appear unfinished, such as the ceilings (deliberately so!), the approach to the large fully-glazed drum enclosing the ticket hall is a powerful and dramatic architectural statement, particularly when seen at night.

Canary Wharf

Designed by Sir Norman Foster and Partners, this is the flagship station on the line, with its cathedral-like volume. As mentioned previously, Paoletti wanted to bring natural light down into his stations wherever possible and this station shows how successfully this goal was achieved. Like most of the stations it was created within a huge excavated box, but the beautiful sculptural

BELOW Canary Wharf – escalator shaft illuminated by natural light. *(Author's collection)*

ABOVE Canary Wharf – the heroic scale, aerofoil sections and sumptuous shapes. *(London Transport Museum)*

qualities of the formed concrete ceiling with its seven aerofoil-section columns belies such a construction. On the platforms are larger interpretations of the classic seating roundels from the 1930s and 1940s, this time fashioned from terrazzo and stainless steel.

North Greenwich

Designed by Alsop & Stormer, this is another huge station, with an equally outstanding adjoining bus garage designed by Foster & Partners. It was originally the gateway to the Millennium Dome, which is now a hugely popular entertainments centre. Paoletti had always wanted one 'blue' station on the line, and this he got in full measure with abundant use of blue mosaic and generous use of backlit cobalt blue glass as a wall cladding.

Canning Town

Designed by John McAslan & Partners, this building is unusual in that it is a triple-decker station: the Docklands Light Railway tracks are 'floated' above those of the Jubilee line, with the ticket hall positioned in turn below it. Perhaps better than on any station, the photograph shows how dramatically the station concourse is illuminated by natural light from above, with views to the platforms overhead.

Stratford Station

Finally, this is another massive station, with a concourse designed by Wilkinson Eyre Partners

RIGHT Canning Town – natural light streaming down. *(London Transport Museum)*

ABOVE Stratford – once again, monumental in scale. *(London Transport Museum)*

and the platforms again by John McAslan. It is more akin to a major international airport terminal building in terms of its space and style. Connections are conveniently made again to the DLR and the Central line, where this time around the passenger walkways are refreshingly clad with white vitreous enamel rather than bare unfinished concrete.

New trains for the Jubilee line

In the spring of 1991 the intention was to fully refurbish the 1983 Stock trains and then build more of the same design so that the extended line could also be adequately served. With this in mind, Warwick Design Consultants were commissioned to develop and produce a half-car-length mock-up. The silver grab-poles and handrails, together with the accents of purple armrests and aquamarine ceiling strip panels and partial door covings, were intended to promote an appropriately regal 'Silver Jubilee' look.

Throughout the uncertain period of the JLE's gestation, it was found that a fleet of brand new trains could be bought at virtually the same price as extensively refurbished ones (including new ones built to match). Therefore a decision was made to do just that, which sounded the death knell of the unpopular 1983/86s, which had therefore all been scrapped by the end of 1998 (when some were only 12 years old).

A contract was awarded at the end of 1999 to GEC Alsthom Metro-Cammell (later to be known as Alsthom SA) to build 59 new six-car trains. The cars themselves were built and painted in Alsthom's Barcelona factory, but the fitting out of the interior, final assembly and testing were carried out at their Washwood Heath plant in Birmingham. This was a good example of today's increasingly international train-building trend, because the doors came from Canada and the rubber suspended bogies were built in France by Alsthom.

Like the 1992 Central line stock, they were again of all-aluminium welded construction. The large panoramic windows of the 1992 trains were unfortunately not carried over, however, because the commission was written around a previously placed contract with Met-Cam that embraced the original plan to perpetuate the same window proportions as the 1983 trains. Nevertheless, an innovative treatment on the car bodies was to use 'ribbon glazing' on all of the passenger windows, in which each set of two flush-fitting passenger window panels is linked together by a black-painted mask to create the impression of continuous window panelling.

Safety criteria had moved on and a different cab was designed, fitted with a similar set of folding detrainment steps as incorporated on the refurbished '73s. Warwick Design Consultants did the engineering design and prototyping and Westinghouse then developed the production version. The folded steps take up much of the available room in the centre cab door; this explains why the central window is so shallow. In order to blend it with the deeper cab windows, these were angled up to achieve a visual link, and the flat, missile-proof glass retained for these again from 1983 Stock.

Another safety factor built into these trains was crash pillars located at the cab corners. These provided the highest level of crash resistance achieved by any tube train, being able to withstand a 100-tonne end load without deformation of the body structure. Warwick also designed the interior of the air-conditioned cab, using established ergonomic criteria to optimise the design and positioning of the controls. The cab console contains two liquid crystal monitors which show images of the platforms, while a third forms part of the train management system.

Rather than start all over again it was decided to use Warwick's original mock-up as the basis for the new interior design. As previously mentioned, this interior had already been used as the inspiration for the 1973 Stock refurbishment, hence the close family resemblance still seen today.

The original design has been progressively diluted over the years. First to go were the lightly brushed stainless steel handles. These were regarded as the worst of all finishes for the visually impaired, and they were replaced by 'grapefruit', a pale yellow recommended by the Loughborough Institute of Consumer Ergonomics. As time went on the dedicated moquette that featured all of the colours of the interior was replaced by a new solution (shared by Northern and Piccadilly line trains), an initiative from Amey, part of the Tube Lines consortium (see page 172) that arose out of the Public-Private Partnership, or PPP. The armrests were also changed to blue, and as a result the carefully orchestrated original 'line identity' design was lost. At the time of writing this moquette has in turn been

BELOW 1996 Jubilee line train with emergency detrainment steps deployed. *(LURS collection)*

replaced by a much more attractive one, 'Barman', which is being standardised for use across the tube fleet.

Four multipurpose areas per car were supplied; these are perch seats, but with a low-slung horizontal handle fitted below that enables wheelchair-bound passengers to conveniently grasp it and orientate themselves into position. Low-level passenger alarms were located for their use in this area, and this same height was followed throughout the car. Four scrolling dot matrix passenger information modules were fitted per car.

Further vehicle developments

Twelve years after plans to install a moving block signalling system for automatic train operation had to be scrapped, a different method was quietly brought into service in June 2011. This system is called transmission-based train control (TBTC), and was developed by Thales UK. It is based on the Alcatel SelTrac system used on the Docklands Light Railway, which is now part of Thales.

Broadly speaking, moving block signalling has distinct advantages over the traditional fixed block system, which relies on track circuits or axle counters to confirm when a train has left a defined length of track (or block) before allowing another train into that area. The number of clear 'blocks' available per train depended on the line speed and braking distances.

By contrast, a moving block system can adjust the separation of trains in real time. It allows them to run closer together and thus reduces train waiting time, whilst ensuring a minimum safe braking distance between trains so that they can brake to a stop before reaching the train in front. Safety is assured by five vehicle control centres (VCCs) on the line, who communicate with one normal and one standby vehicle on-board computer (VOBC) in each train, and a failsafe design ensures that on failure of a VCC or VOBC or any other loss of communication the brakes are automatically engaged.

Another development was the procurement of an additional seventh car per train. In 2005 it was announced that London Underground would add a seventh car to the 59 sets of 1996 Stock plus four new trains. These were completely sourced from a new factory that Alsthom had built in the suburbs of Barcelona, and of course they form a seamless addition to the existing trains.

ABOVE Interior when new with original moquette design. *(Wikimedia Commons – Chris McKenna Thryduulf)*

New trains for the Northern line

Known as 1995 Stock, these are almost identical in visual terms to the Jubilee line trains, but irrespective of their date are actually a more modern design with some fundamental technical differences, the reason being that, as has already been explained, the design and technical specification of the '96 was signed off back in 1991. The latter were designed for 'cheapest first cost' for London Underground, while the 1995 Stock was designed for 'lifecycle cost' because Alsthom have a stake in them, having won the contract to service and maintain the fleet through an Alsthom company called Northern Line Service Provision Ltd.

These new trains had been urgently required for the Northern line, which was still languishing under the 'Misery line' epithet bestowed upon it by the *Evening Standard*. The Northern got them via a service provision package, worth at the time £40–£45 million a year. Under this leasing deal, Alsthom supplied the trains under a PFI (private finance initiative), and apart from running the line's maintenance depots they would also be responsible for providing trains

ABOVE External appearance when new. *(Giles Barnard)*

for service each day over a 20-year period. The entire cost of the trains was arranged by Alsthom using an operating lease – two major banks bought the trains for London Underground, which were then leased back over an agreed timescale, offering funding and cashflow benefits.

Key design differences from previous stock are as follows:

- Adtranz air-suspended bogies were used rather than rubber, to cope with the demanding track conditions found on part of the line.
- The cab's centre door folds inwards, and unlike '96 Stock a separate fold-out set of steps is deployed for emergency egress of passengers. This creates a deeper window that more effectively aligns with the shape of the other cab windows.
- A more modern AC traction control system is used, being Alsthom's 'Onix' three-phase insulated gate bipolar transistor drive. 1996 Stock, on the other hand, specified three-phase induction motors fed from a single source inverter using a gate turn-off thyristor.
- The interior was modified to give a line identity; therefore black armrests and passenger information module surrounds were finished in black and the 'grapefruit' pale yellow colour for handles and grab-poles was carried over. Tip-up seats replaced the large perch seat in the multi-purpose area.
- The dedicated art deco-inspired moquette design that was first fitted was subsequently replaced first by the Amey solution and now by the 'Barman' design.

A half-life refurbishment of the interiors currently under way not only features this moquette but also removes all of the yellow grab-poles, replacing them with corporate blue ones throughout. The overall look is certainly smarter and more distinguished and one can only assume that the partially sighted are still happy with this change.

BELOW Grab-poles change from yellow to corporate blue. *(Mike Ashworth/London) Underground*

Chapter 14

The Millennium years (2000 to date)

This brief review covers the major political changes that affected the London Underground during this period.

Creation of 'Transport for London' (TfL)

This body was established in 2000 as the new transport authority for London. Part of the Greater London Assembly, it gained most of its functions from the earlier London Regional Transport, but for reasons outlined later it did not take over responsibility for the Underground until 2003.

Development of the Public-Private Partnership (PPP)

This proved to be the most dramatic restructuring of the Underground throughout its long history, and brought with it an unprecedented amount of political posturing, infighting and eventual failure.

The substantial Jubilee line extension overspend had convinced both the Treasury and the incoming Labour Government (swept into office in 1997) that London Underground was incapable of delivering major projects to time and on budget. The new government was committed to modernising public infrastructure, such as transport, hospitals and schools, without adding to the public sector borrowing requirement (PSBR). They consequently sought to employ the allegedly more efficient and superior project management and delivery skills to be found in the private sector through private finance initiatives.

As far as the Underground was concerned, private investment now and repayment from income over the created assets' lifespan seemed to provide the long-term investment needed for their major projects. In 1997, a PPP seemed to be the best way forward, and it was supported by the Underground's senior management, for the company was then haemorrhaging cash in a big way. A major attraction was that a PPP could create stable and increased investment on a rolling basis to ensure that capital projects would be adequately funded throughout their life until completion. A further incentive was that an investment backlog of £2 billion could then be tackled, which would in turn signal the end of the outmoded practice whereby the Underground had to present an annual bid to the Department of Transport in order to secure their monies for just the following fiscal year, with no assurance that a major project could be completed.

As Minister for the Environment, Transport and the Regions, John Prescott was the man responsible for taking the PPP forward (as well as for creating the Greater London Assembly and the Mayoral Office), and its creation was announced in March 1998. The publicly-owned element (London Underground) would be responsible for the operation of the train service, which included running the trains and stations, setting fares and controlling safety issues and signalling. The private sector would be responsible for the maintenance and renewal of the infrastructure (rolling stock, stations, tracks, tunnels and signals). The contracts were initially to be for 15 years and were worth £7 billion, with London Underground bearing none of the cost.

In September 2001 the two successful syndicates were announced: Metronet Rail Ltd (composed of W.S. Atkins, Bombardier, EDF Energy, Thames Water and Balfour Beatty) won the contract for all of the four sub-surface lines as well as the Bakerloo, Central and Victoria, while Tube Lines (composed of Jarvis, Amey and Bechtel) was awarded the Jubilee, Northern and Piccadilly franchises.

Over the next two years its gestation path became extremely convoluted by much internecine wrangling, particularly between the then new mayor, Ken Livingstone (who sought legal challenges that were overridden), Transport for London and the Government. As a result, by 2003 it was already judged to have become fatally flawed and diluted. In that year it was announced that the contracts would be worth £13 billion over 30 years and would now cost London Underground £1 billion a year.

For all sorts of reasons – such as late and under-delivery of contracts and labyrinthine financial complexities that were virtually impossible to master – costs spiralled out of control, and Metronet was forced into receivership in July 2007. In order to keep the business activities going it was bailed out by the UK Government at a cost of £2 billion, and in 2008 the contracts and employees were transferred to Transport for London. In December 2009 the business finally reverted back to London Underground's control.

By contrast Tube Lines had brought in almost all of its projects on time and to budget, as they placed work out to tender rather than employing the shareholders as suppliers, and achieved better cost control. However, after heated debates between them and London Underground on the cost of works for the second seven-and-a-half-year contractual period (which an independent arbitrator had failed to resolve), TfL finally agreed to buy the shares of Bechtel and Amey (Ferrovial) from Tube Lines in May 2010 for £310 million; this amount was happily recovered in two years by cheaper debt financing.

Despite all the financial and political shenanigans, the privatisation of maintenance and renewals did deliver some very positive results during this period, such as improvements in service and enhanced standards of station design (such as the new Wembley Park), as well as a noticeably cleaner system. And of course, excellent, innovative new trains designed to very high standards were also developed, which will be described in the following section.

The new Bombardier 'Movia' trains

New 2009 Stock tube trains for the Victoria line

An order was placed with Bombardier Transportation for 47 eight-car units to be built

BELOW 2009 Tube Stock and 2011 Surface Stock side by side. *(Jon Ratcliffe)*

at their Litchurch Lane, Derby, works. This was part of a £3.4-billion contract awarded by Metronet Rail Ltd to supply new trains with ATO and signalling for the Victoria line as well as other sub-surface lines. These entered service from July 2009 onwards, finally displacing all of the 1967 Stock by June 2011. Each train is made up of two four-car units as follows: driving motor, trailer, non-driving motor and an uncoupling non-driving motor (used for shunting purposes).

Part of Bombardier's 'Movia' family, the trains continued to be built by means of the now common practice of sub-assemblies being made from welded aluminium extrusions which are then 'huck-bolted' to form a shell. London Underground elected to continue with this manufacturing technique, although Bombardier has developed another manufacturing technology for metro vehicles. This is called FICAS (fully integrated car-body assembly system) and embraces an innovative modular sandwich construction that comprises stainless steel panels bonded to a rigid foam core, an interesting crossover of technology gained from their aerospace division. This can result in a thinner body cross-section that increases the interior passenger capacity by up to 10% without impinging on the exterior dimensions. Other significant gains are reduced vehicle weight, which means less power consumption and reduced track pressure, as well as a superior external finish. These benefits do not detract from the latest crash-resistance standards and required insulation levels. However, although the first FICAS trains went into service on the Stockholm Metro in 2003, the technology's time had clearly not yet come for London Underground.

The 2009 Stock trains are faster than those that they replaced, having a higher top speed of 50mph, and at peak times 43 trains should be in service, an increase of six over 1967 Stock. They are now the longest deep-level tube trains running on the network, being 9.8ft longer than their predecessors. The exterior cab styling certainly took a new direction from all previous designs, being more rounded and softer in shape, which is accentuated by curved glass cab windows. A prominent 'peak' formed on top of the central cab door lends visual prominence to the illuminated destination display. The interior design is a pleasing evolutionary step forward from its predecessor, with particularly careful detailing of all its forms; the seat backs are, however, harder!

The 'ivory' shell colour remains as a corporate solution, and the moquette seating fabric is a very close cousin to the one designed by Tilney Shane 20 years earlier. The elegantly formed cast aluminium ivory armrests are also in-filled with plain blue moquette. The integration and performance of the luminaires is very good and builds on all the experience gained since the late 1980s.

Also impressive is the scooped-out sculptured detailing of the portal mouldings around the doors, allowing grab-poles in the line colour to be fully integrated. All other vertical grab-poles (including the slightly curved ones, a first for Underground trains) are in the darker corporate blue, with all horizontal handrails in the lighter blue shade. Colour contrast is important for the partially sighted, particularly in the door areas, so perhaps the colours should have been reversed in this regard.

Wheelchair-bound passengers are particularly well catered for in these trains, as they were the first Underground trains to be designed since the Rail Vehicle Accessibility Regulations were initiated. In a designated multipurpose

BELOW Interior of 2009 Stock. *(Author's collection)*

ABOVE The cancelled real-time information screens. *(The Transport Design Company)*

car, an area is provided with six tip-up seats for wheelchair users and pushchairs. These are separated in their centre by a backboard of prescribed dimensions with its own offset curved grab-pole to aid wheelchair manoeuvrability. All of the Victoria line stations (except Pimlico) have a locally raised platform area for easy entry into these cars.

A feature that was originally planned for these trains would have been the provision of thin film transistor (TFT) or LCD screens in the car. These are now commonplace in European trains (including mountain trains in Bavaria!), and give a host of real-time messages and tourist information in crisp, clear colour as illustrated. Designed by the Transport Design Company, two screens would have typically shown service disruption and interchange information (currently only given audibly) and appropriate messages with pictograms that recognised London's diverse passengers, and in an emergency the background colour would change to yellow to reinforce safety messages. Unfortunately, however, cost ruled out this most useful passenger-orientated feature, and the now standardised scrolling dot matrix message systems were fitted instead.

2011 'S' Stock-type Surface Stock trains

These are also part of Bombardier's 'Movia' family, and in either S7 or S8 form will eventually serve all of London Underground's sub-service lines. This is the first time in its entire history that an essentially 'one size fits all' design will be used in this way. Only the amount and types of seats and the number of cars per train will change to suit a line's requirements; thus the S8 cars form an eight-car train for the Metropolitan line and the S7 cars will make up seven-car trains for the rest.

A standardised fleet of 191 trains will be made up from 1,395 cars at a cost of £1.5 billion; the order is quoted as being the largest single order for trains ever awarded in Britain. They first entered service on the Metropolitan line in July 2010, the Hammersmith & City in July 2012 and the Circle from September 2013, and they have already been seen on the District line.

They have several features new to London Underground trains, including:

- Air-conditioning, with units supplied by Mitsubishi. Two circuits are provided so that if one fails, 50% of performance capacity is retained. This equipment is appropriate for use in sub-surface tunnels because the exhausted hot air can be readily dispersed (unlike in tube tunnels), and two-thirds of the sub-surface network in any case runs in the open air.
- Full open access from car to car throughout the train, with 1,250mm clearance at floor level and 1,700mm clear space overall. This feature had just been introduced to Londoners on the excellent new 'London

RIGHT **Interior of 'S' Stock, Metropolitan line version.** *(Wikimedia Commons – Spmiller)*

Overground' trains, also built by Bombardier. The Hong Kong Metro trains had this feature from the start, and Berlin introduced it some 20 years ago. Since the same Rail Vehicle Accessibility Regulations had to be met, a similar designated multi-use car makes up a train, with six tip-up seats and a central backboard.

- When fitted in a Metropolitan line train, the 16 transverse seats per car are cantilevered completely clear of the body sides (another Underground first) while the under-seat equipment cabinets are fully recessed under the longitudinal seats, except for those fitted either side of the connection concertina, where they are just underflushed.

The external cab styling with its curved glass windows (another first for a sub-surface train) has a strong family resemblance to the 2009 Stock, not surprising since both were designed by Bombardier's own in-house industrial design team. This arrangement clearly works very well, because team members are always on hand to work through problems with engineers and clients to resolve emerging difficulties as the design develops. In this way they are able to monitor an acceptable transition of their design

ABOVE **The multipurpose area showing the 'backboard' as also fitted in the 2009 Stock cars.** *(Wikimedia Commons – Spmiller)*

LEFT **The articulated section.** *(Wikimedia Commons – Hahifuheho)*

concepts through to production; the results speak for themselves.

The cars feature a pronounced and attractive tumblehome from waist level to sole-bar, which lends distinction to their design.

DCA Design Consultants were commissioned to design the emergency detrainment folding steps built behind the cab's centre door, another first for any Surface Stock train. These are a lightweight design that also incorporates an interlocking bridge plate, which not only clears the auto-coupler but also allows train-to-train evacuation. It can be deployed in under a minute, which thanks to the train's open car access allows the speedy evacuation of 1,800 passengers at the rate of one every two seconds.

It had been decided to standardise on yellow grab-poles and handrails throughout the entire fleet for all four lines to ease production and procurement. A number of yellow strap hangers (as used on Bombardiers Class 378 'London Overground' trains) have since found their way back on to these handrails after a long absence from the Underground. Refurbished 'A' Stock trains had horizontal handrails set at a height that seemingly elicited no complaints, but passengers have found the ones on 'S' Stock to be uncomfortably high, even though the population is supposed to be growing taller!

Ticketing system developments and improvements

The UTS was dramatically improved from June 2003 onwards by the introduction of Oyster, a stored value electronic 'smart card' system which is actuated on gates and ticket machines by swiping a yellow pad. (This had been set up as another PFI contract between TfL and a consortium of suppliers that included EDS and Cubic Transportation Systems.) Its main attraction is the ability to 'pay as you go' by loading a given cash value at the point of purchase (on-line sales, ticket offices and about 4,000 designated retail outlets at the time of writing), which can then be topped up at the passenger's convenience. By the end of 2012 over 43 million Oyster cards had been issued and were being used for more than

RIGHT Detrainment in action. *(DCA Design International)*

80% of Underground, bus, tram and London Overground journeys.

An additional refinement was the introduction in 2007 of a credit card variant by Barclaycard called 'One Pulse', which combines Oyster Card functionality with Visa credit card facilities.

A significant post-Jubilee line extension station

King's Cross/St Pancras Underground station has been dramatically redesigned in recent years and is the most stunning example of 'before and after' yet seen on the system. The previous station was just too small, cramped and completely out of date with its network of dreary labyrinthine rat runs leading to six different tube and surface Underground lines. Furthermore, interchange to all of these had often to be achieved by the irksome requirement of having to pass through a sequence of ticket gates.

Thanks to an overall spend of £810 million funded by the Department of Transport (with a contribution of €20 million from the EU Trans-European network) all of this has now been swept away by the creation of two new expansive ticket halls that are everything that its predecessors were not. They are bright, cheerful and welcoming, and the generous well-lit spaces are now more typical of an international air terminal than an Underground station. (This impression is heightened by the number of passengers passing through towing wheeled cases!)

A particularly important factor for every type of passenger is that step-free access is now provided to all six lines. The longevity and timelessness of the HLS signing design, now 30 years old since concept work started, still proves to be absolutely up to date and can meet any operational challenge, as the accompanying photograph demonstrates.

Three hundred thousand passengers use the station daily, and capacity has been quadrupled by the new ticket halls to meet demand that can only ever increase. Currently it already deals with more passengers than Heathrow Airport! The whole area around King's Cross is being extensively regenerated and the ever-increasing international traffic from the neighbouring St Pancras International will also swell numbers considerably.

BELOW Plenty of space in the new King's Cross/ St Pancras station complex. *(Copyright Thomas Graham/Arup)*

The new Underground station now blends seamlessly with both the adjacent mainline stations, which have also been transformed in recent years to meet the highest international standards of railway station design. The Mayor of London, Boris Johnson, quite rightly said on the opening of the second Northern ticket hall in 2009 that 'It is the standard by which all new station developments should be judged', and how right he is! The success of this complex must surely point the way for future large-scale redevelopments of busy stations such as Victoria and Bank.

Some notes on maintenance and operation

Power supply and generation

The original Lots Road 'Chelsea Monster' that eventually powered most of the rail lines and tramways within the Underground Group became operational in February 1905 and was only finally decommissioned in October 2002, having been in use for almost a hundred years. Over this period it had been refitted several times. Initially it burnt 700 tonnes of coal a day with a generating capacity of 50,000kW. Modernisation in the 1960s converted the station to 50Hz generation, with coal changing to heavy fuel oil. In the 1970s, following the discovery of North Sea gas, the boilers were converted to burn this fuel but retained the option to revert to oil if required. During this time the Underground network received all is power from Lots Road as well as from an ex-LCC Tramways power station in Greenwich.

In the 1990s a decision was made not to update and re-equip the station and to henceforth purchase generated power from the National Grid via EDF Energy.

As for Lots Road, after several redevelopment stops and starts, Boris Johnson broke ground in September 2013 on an eight-acre site which will be called the Chelsea Waterfront, a £1 billion regeneration which promises to be the biggest riverside development on the north side of the Thames. Phase 2 is to be completed in 2017/18 and will include the old power station building, which will be converted into shops, restaurants and flats.

Track in tube tunnels

The London Underground has always used the old UK standard bullhead rail section weighing 95lb per yard, but in recent times there has been a determined effort to convert to flat-bottom rail as used elsewhere around the world.

All lines operate at 630Vdc and use the characteristic four-rail system. The negative rail is positioned 1.5in higher than the running rails and is mounted centrally between them. The positive rail is located 3in higher and outside the running rails, away from the platform side of the track, its standard position usually being on the left-hand side of the direction of travel. The two increased heights allow the collector shoes fitted on the trains to comfortably clear the running rails.

In deep tube tunnels and stations the current rails are of lighter section than the running rails (which have by now all been converted to flat-bottom rail) and use white porcelain pots for their insulation. In these environments, the rails are mounted on small wooden blocks set in concrete. In the tunnels a shallow pit is provided between the rails that is filled with ballast to provide both sound deadening and drainage. In the stations the familiar 'suicide pit' is provided.

Tracks and track maintenance

As can be imagined, rail replacement in these parts of the system is a particularly slow and arduous process, since the blocks have to be drilled out and the concrete renewed. This is aggravated by the fact that there are typically only four hours available at night for all track and lineside maintenance to be carried out before the first trains start running again. Battery locomotives to a more or less standard design have been used for many years to haul works wagons carrying a variety of required materials and equipment to carry out all of this work during the limited time available.

On the sub-surface lines and outdoor sections concrete sleepers are used, and here the limestone ballast is being replaced by granite, initially more expensive but cost-effective over time. Much of this work is locally based but materials and heavy machinery are stored at the Ruislip and Lillie Bridge depots.

Tunnel cleaning

Clearing the tunnels of dirt and dust has always been a major problem for the Underground. This residue is primarily iron-based from the wheel/rail interface, and is blended with oil and grease from lubricants and, of course, trace amounts of human hair and skin! This material can be mistaken for smoke when it becomes airborne and can trigger allegations of fire incidents as it can smoulder when heated.

Over the decades tunnel cleaning was

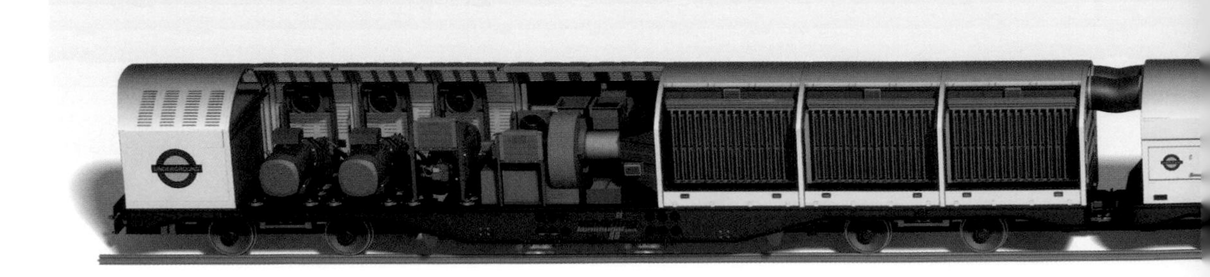

achieved by hand; first of all by teams of women, called 'Fluffers' who descended into the tunnels by night armed with just dust pans and brushes. Even today contractors are retained to clean tunnels manually overnight using vacuum cleaners, but it would take two years for six gangs of six men to deal with all the tunnels, so large numbers of teams are required to clean the network every year. The tunnels, at 220km represent 55% of the 402km network!

In 1976 a tunnel-cleaning train entered service, sandwiched between two cars converted from 1938 Stock vehicles. It used forced air to disturb the dust followed by a vacuum to remove it. It was a worthwhile attempt to deal with the problem but fell short of optimum performance requirements in several areas:

- It required multiple passes to clean a single section of tunnel.
- It caused dust clouds when it succeeded in disturbing dust, but often failed to capture it.
- It had several blind spots, and could not clean the tunnel roof.
- Its operation was spasmodic, working on an 'as required' basis.

In the end it was taken out of operation due to unreliability and high maintenance and operating costs. It has been out of operation since 2002 and has now been scrapped.

A brand new tunnel cleaning train

The two-year period to clean the whole network of tunnels by traditional methods will now be dramatically slashed to two months by the introduction of a new state-of-the-art train. This will consist of seven cars headed and tailed by a pair of coupled ex-1967 Victoria line driving motor cars. These are completely modified to provide traction and to supply power for the vacuum and dust disposal plant built into the three centre vehicles (built by Schörling Kommunal, world leaders in the supply of metro cleaning equipment). The two outer frame-mounted vehicles each carry a control system, an engine compartment, a suction fan compartment and a layered fine-dust filter compartment.

The shorter vehicle at the centre has compressed air and vacuum nozzles at its centre with litter and coarse material filters either side. A 360° ring of 13 nozzles, looking exactly like large domestic vacuum cleaner nozzles, are mounted around its periphery to ensure that all the dust is captured. These can be spread to optimise cleaning within a range of tunnel bores so that no blind spots are left and controlled jets of air are produced rather than a slow moving body. All of the dust is then collected with two opposing vacuum streams. The mainly double-track sub-surface tunnels can also be cleaned.

Just one pass is adequate, which means lower site access requirements. A hydraulic drive allows low-speed running at 1kph during cleaning, while the motored cars allow the train to reach its required destinations at line speed.

The Asset Inspection train

This is another specialised six-car train that is currently making its appearance. Its brand new appearance belies the fact that it is, in fact, made up from substantially modified old vehicles. These are an ex-Northern line 1972 Mark 1 car at each end and two ex-Victoria line 1967 driving motors in the middle.

Commissioned by Tube Lines, it replaces

BELOW The new tunnel-cleaning train. *(Schörling Kommunal GmbH)*

BELOW The Asset Inspection train alongside the older track-recording vehicle. *(Brian Hardy)*

the existing track recording vehicle (which has 1960 Tube Stock pilot cars and a 1973 Tube Stock track recording car) and uses the very latest dynamic technology to check track geometry and rail profiles at high speeds. Thermal cameras are used to monitor cables for hot spots, and high-quality digital equipment feeds back real-time data to the on-board computers for analysis. It will act as a platform for testing future measurement systems (such as enhancements to the automated track measurement system on passenger trains) and integrating ATP systems used across the Underground network. It will also act as a calibration tool and hot standby vehicle for the ATMS equipped trains. It is painted in Tube Line's colours with yellow warning panels on the outer cabs to permit operation on Network Rail infrastructure.

The train can capture data every one tenth of a metre of track, which is recorded through both video and high-resolution thermal imagery that is then collated online. Other technologies include ride quality noise measurement and collector shoe impact detection.

The introduction of all of this 21st-century technology importantly provides the knowledge and ability to plan more cost-effective maintenance regimes based on the infrastructure's requirements, and not on a predetermined life expectancy.

The Underground's two 'finest hours'

The first: the 7/7/2005 terrorist attacks

Ironically, the very day after London won its bid to host the 2012 Olympic Games on 6 July 2005, the Underground railway was subjected to a series of the most despicable and pernicious attacks by four home-grown radicalised Islamist terrorists who were happy to die for their distorted convictions. Three trains with their passengers were torn asunder during their morning commute to work. In all, 52 innocent passengers (including 13 on a bus at Russell Square) were murdered and a further 700 injured in the first suicide attacks to have ever taken place on British soil.

From 8:50am three bombs were detonated within 50 seconds of each other. The first device exploded in a 'C' Stock train travelling eastwards between Liverpool Street and Aldgate, killing seven people. The second, also in a 'C' Stock train, exploded just after it had left Edgware Road travelling towards Paddington, killing a further six.

A Piccadilly line train travelling southwards from King's Cross/St Pancras was just 500yd (450m) into the tunnel when the third bomb exploded. The narrow confines of the tunnel concentrated the effect of the blast and thus the carnage was particularly terrible, resulting in 26 deaths.

Both Underground staff and the emergency services performed magnificently that day at all levels. They demonstrated by their professionalism that the hard-won lessons that had been acquired 18 years previously following the King's Cross disaster could now be applied impeccably and seamlessly in terms of the thorough training that had been given and the rapid response procedures that had been learnt.

Tim O'Toole, the Underground's American then managing director, received an honorary CBE for leading the system's overall response, and Peter Hendy, then MD for surface transport (since knighted), received the same accolade. But the bravery of men and women on the ground was also fully recognised. There were many unsung heroes that day, but in particular MBEs were awarded to:

- David Boyce, Russell Square station supervisor who ran into the tunnel to give first aid to the victims, many terribly injured.
- John Boyle, an off-duty train operator who ran into Aldgate Station to evacuate passengers.
- Tim Wade, East London line manager who entered the Piccadilly line tunnel to assist the wounded.
- Peter Sanders, station manager at King's Cross/St Pancras Underground station.

In addition to these London Underground personnel, Alan Dell from London Buses was also awarded an MBE, and a further group of 19 CBEs, OBEs and MBEs were presented to various echelons of the emergency and security services as well as staff from local hospitals.

The second: the 2012 Olympic Games

The London Underground met with flying colours the supreme challenge of having to deliver a safe and speedy travelling experience for the huge influx of passengers who used the system to reach all of the 2012 Olympic events across Greater London and elsewhere. A system of bold temporary graphics with on a magenta background worked admirably in the achievement of this goal, and the immense scale of the Jubilee line's Stratford station was an enormous bonus in being able to marshal the huge crowds on their way to the Olympic Village.

The 150th year celebration run

On 13 January 2013 a special steam-hauled train re-enacted the very first day's service on the Metropolitan line by running along the original route almost 150 years to the day. The locomotive was Metropolitan No 1, dating from 1898 (described in an earlier chapter), which hauled a rake of historic Metropolitan wooden coaches. Amongst these was number 353, sole survivor of the first-class 'Jubilee' four-wheelers from 1892, that had emerged from an extensive two-year restoration.

The photograph of this train at Baker Street station surrounded by smoke and steam makes a fascinating comparison to C. Hamilton Ellis' painting illustrated at the beginning of this book showing Daniel Gooch's broad gauge train on that first day, also at Baker Street.

Where we came in then...but not quite!

BELOW Where we came in, but not quite! *(Transport for London)*

Epilogue

The future

New extensions

1. The Croxley Rail Link

The Metropolitan line will finally, via a new station at Watford High Street, have a direct connection with Watford Junction on the West Coast Main Line, as well as London overground services. The location of the original Watford terminus building, built in 1924 (another of C. W. Clark's rural 'vernacular' designs), had never been ideal, since it was situated approximately one mile from the town centre.

The government approved the circa £118 million project in July 2013, and the re-routed line is due to open in 2017. Two new stations will be built en route – Cassiobridge and Watford Vicarage Road, which will serve the local hospital and football stadium.

2. Proposed Northern line extension from Kennington to Battersea

Subject to funding and permission to build and operate the extension, two new stations at Nine Elms and Battersea could be open by 2020. In terms of benefits, the extension will improve transport links and public spaces in the area, and is deemed essential to support the transformation of Vauxhall, Nine Elms and Battersea, a designated regeneration area on the South Bank of the Thames. It is estimated that up to 25,000 jobs and 16,000 new homes could be created, and journey times from the two new stations to the West End and the City could be, in some cases, less than 15 minutes.

A further order of trains, compatible with existing Northern line stock, will be required to deliver these new services, possibly with enhanced or different traction equipment.

BELOW Blackfriars Station, an example of the latest architectural style. The station`s services are shared with National Rail. *(Author's collection)*

The challenges ahead

Over the last decade, the London Underground system has seen an ever-increasing demand for its passenger journeys over the available service volume supply. A reliable and efficient service is essential to meet this need, and the last ten years has already seen a 56% improvement in reliability. In order to respond to such pressures, and the changing requirements of London as a world city, a number of major station upgrades are either at the planning stage, or at the early stages of construction. Among these are:

1. Victoria Underground station

This station is currently used by 82 million people per year, and is heavily congested. The congestion is so acute, that at peak times passengers are regularly held for several minutes behind the gate line, or even outside the station, to allow platforms to clear.

As an example of the ever-continuing growth and demand on the Underground's services, passenger numbers at this station are set to increase by a further 20 per cent, and consequently a £700 million project is now underway to enlarge the station area by 50 per cent. Apart from reducing congestion and delays during peak hours, the modifications will provide step-free access between platforms and the street for passengers with heavy luggage or restricted mobility. Construction is underway, and is scheduled for completion by 2018, with a new north ticket hall and an enlarged south one. New lifts, and nine new escalators, will provide easier access from street level to the platforms. There will also be new and more-convenient interchanges between the Victoria and District & Circle line platforms. Expect a similar spacious and welcoming environment to the one already enjoyed at King's Cross/St Pancras.

ABOVE The new station entrance on Marylebone Lane. The fascia design is the latest standard solution. *(Image reproduced by permission of London Underground)*

2. Bond Street and its interface with Crossrail

This station is undergoing a major extension to increase capacity and improve accessibility by 2017. It will also link with Crossrail – the new high-capacity, high-frequency railway that will run from Maidenhead and Heathrow in the west to Shenfield and Abbey Wood in the east.

More than 155,000 passengers currently use Bond Street station daily, and this figure is set to rise to 225,000 when Crossrail arrives in 2018.

There will be a new station entrance on Marylebone Lane, on the north side of Oxford Street, leading to a new ticket hall. New escalators will serve the Jubilee line, and once again lifts will provide step-free access from street to platforms. Interchange to the Crossrail platforms will be made within the station. Crossrail is building two new ticket halls south of Oxford Street, at Hanover Square and Davies Street.

3. Tottenham Court Road and its interface with Crossrail

Tottenham Court Road is another very busy station that currently suffers from acute congestion. For example, around 70 members of staff have to manage over 150,000 passengers a day, and severe overcrowding often creates brief closures during the hours of peak travel. Historical precedent has resulted in the legacy of both a very cramped ticket hall, and street accessed entrances/exits of ungenerous width. Originally, there were two separate stations serving the Central London and the Northern line's predecessor, that were linked by subway in 1908, and then combined into one station following work that took place in 1925. All traces of the original 1900 CLR surface building by Harry Measures were finally demolished in 2009, to make room for the new station complex.

The upgrade consists of two projects – the London Underground congestion relief, and the Crossrail link.

Once the interchange with the Crossrail station is completed on the site in 2018, more than 200,000 passengers will pass through a full refurbishment of existing facilities. These embrace a new and much larger ticket hall with improved access to the existing Underground lines, via step-free access and new escalators. The existing entry and exit routes to the

BELOW The twin fully glazed entrances/exits of the new Tottenham Court Road plaza. *(Image reproduced by permission of London Underground)*

ABOVE Tottenham Court Road – escalators leading down to a generously sized ticket hall. *(Image reproduced by permission of London Underground)*

Northern line platforms are currently very poor – passengers have to enter and leave by one end of each platform, creating conflicting passenger flows. A new layout will resolve this irritating problem by providing access to both ends of each platform.

This is a £1 billion upgrade (£500 million for the London Underground station upgrade, and £500 million for the Crossrail station). The most visible external feature will be the new wedge-shaped fully glazed station entrances, 17 metres tall and set in a newly modelled 'plaza' built in front of the Centrepoint building. Great care is being taken to retain the famous 'pop-art' mosaics by Eduardo Paolozzi (see page 139). These will be complemented by new artwork by Daniel Buren that will grace the new ticket hall sitting under the forecourt of the 1960s tower block which dominates this area.

A modern, cost efficient customer service that harnesses the full potential of modern technology

1. Ticketing The Underground will adopt the technology used by other major retailers, selling its journeys not only with Oyster cards, but also with contactless payment cards, be they credit, debit or charge cards. These payment options have the added benefit of no requirement to top up or buy credit in advance, and can be used for 'touch-in and go' transactions. These payment methods were introduced in early 2014 to meet the Underground management's aspirations for simpler ticketing.

2. More visible staff at every station. With paper/card ticketing rapidly becoming a thing of the past, displaced ticket-office personnel will be able to provide an important customer-service role. They will also be more able to assist passengers face to face with advice and information on journey options, ticketing and service status.

3. Integrated, reliable multi-modal passenger information in real time. At some point this will be available within the trains, but will be repeated and reinforced by the same consistent messages on static station information displays, and also on the passengers' own smart-phones and hand-held devices.

4. Maximising commercial revenue. There is no bottomless pit from central government to fund all of these exciting schemes, and money must be found elsewhere. Therefore, taking a leaf out of Network Rail's book, the Underground wishes to attract retail outlets of the highest quality and calibre, particularly to enhance all of those large, bright new spaces it is creating.

5. Third party financial involvement in new projects. The Underground will seek more financial partners by attracting businesses to share development costs. The Piccadilly line's extension to Heathrow Airport Terminal 5 is a prime example of such a strategy. Opening in the spring of 2008, the British Airports Authority co-ordinated the Underground extension design with the rest of the terminal project and paid for it!

Train service and performance

From 2015, for the first time there will be a 24-hour service on Fridays and Saturdays. The initial 24-hour weekend Tube network will comprise regular services on the Northern, Piccadilly, Victoria, Central and Jubilee lines.

Following on from the Victoria line trains of 1967, which were the first in the world with Automatic Train Operation (ATO), the Central line converted to ATO in the mid 1990s. Here, the trains are computer controlled – the driver opens and closes the doors, monitors the train and fixes faults on board where possible. Other lines have since followed suit.

The Jubilee was converted to ATO in 2011, and the Northern line will complete the switchover later in 2014, after the first section was finished in 2013. The driver operates the doors, and presses the buttons to start the train, and if the train stops for 'fail-safe' reasons, he or she then takes over. These measures dramatically assist in the delivery of greater capacity.

Peak capacity on the Northern line is planned to increase by 20 per cent by 2014, providing 24 to 30 trains per hour. It is planned for the Victoria line to increase services from the current 33 trains per hour to 36 by 2016, and the Jubilee line from 30 to 36 trains per hour by 2020. The latter increase will require additional rolling stock for the line.

The reason that these improvements are possible is that since the trains are controlled electronically, acceleration and braking curves, and therefore performance, can be continually optimised.

Future deep-level tube trains for the Piccadilly, Central, Bakerloo and Waterloo & City lines

On 28 February 2014, London Underground unveiled its plans to buy 250 new-generation trains for these lines, which will be designed to serve for 40–50 years. These plans will be supported by a sustainable programme of investment across the four lines, with resulting economies of scale. A notice was placed with the Official Journal of the European Union, seeking expressions of interest from the rail industry to build the trains. A formal Invitation to Tender is expected in early 2015.

The new trains may be able to run without drivers, as they would be 'capable of full automation' in every respect. The trains would also be able monitor their own health using on-board diagnostic equipment and transmit this data back to base.

The requirements stipulate that the new trains will be more reliable, will have air-conditioning installed and will have a similar 'walk-through' configuration to 'S'-Stock trains. The accompanying illustration shows how these new trains may look, but it will be up to industry, working closely with London Underground, to thrash out the final design and specification.

It is more than likely that the latest new trains from Bombardier will be the last to be ordered with conventional cabs, apart from those 'clone' trains for the Northern and Jubilee extensions, which will remain conventional.

Because of a future 'migratory' phase, during which the new trains will have to mix with the old in service, a manual driving position could be provided, albeit that the driver need not be able to see out of the cab and could be relayed the view outside via screen images. The Helsinki Metro has ordered new trains that are cabbed, with the proviso that they can be converted to driverless at a later date when the full operating technology is installed. Indeed, as this technology enters its mature phase, driverless trains are being increasingly specified for metro systems around the world. For example, the Paris Metro has recently converted to driverless trains to operate its original 'Line One' route.

To sum up, the future challenges are particularly exciting, as the Underground is poised to encompass the very latest technology in concert with respect for its unique design heritage to provide an outstanding world-class service. Heady days indeed!

ABOVE Tottenham Court Road concourse area. *(Internal cladding design copyright Daniel Buren: image reproduced by permission of London Underground)*

BELOW One of the design proposals currently being considered for the Deep Tube project. *(Image reproduced by permission of and copyright of London Underground 2014)*

Glossary of acronyms

AFC – Automatic fare collection.
AIT – Asset Inspection Train.
ATMS – Automated track measurement system.
ATO – Automatic train operation.
ATP – Automatic Train Protection.
BTC – British Transport Commission.
BTH – British Thomson-Houston.
CLR – Central London Railway.
C&SLR – City & South London Railway.
DIA – Design and Industries Association.
DLR – Docklands Light Railway.
DRU – Design Research Unit.
FICAS – Fully integrated car-body assembly system.
GLC – Greater London Council.
GN&CR – Great Northern & City Railway.
GNR – Great Northern Railway.
GTO – Gate Turn-off.
GWR – Great Western Railway.
HLS – Henrion, Ludlow and Schmidt.
JLE – Jubilee line extension.
LCC – London County Council.
LER – London Electric Railway.
LGOC – London General Omnibus Company.
LNER – London & North Eastern Railway.
LNWR – London & North Western Railway.
LPTB – London Passenger Transport Board.
LRT – London Regional Transport.
LSWR – London & South Western Railway.
LT – London Transport.
MDET – District Electric Traction Company.
MRCE – Metropolitan Country Estates Ltd.
NATM – New Austrian tunnelling method.
OPO – One-person-operation.
PFI – Private finance initiative.
PPP – Public-Private Partnership.
TBTC – Transmission-based train control.
TfL – Transport for London.
UERL – Underground Electric Railway Company of London.
UTS – Underground Ticketing System.
VCC – Vehicle control centre.
VOBC – Vehicle on-board computer.
WCL – Westinghouse-Cubic Limited.

Bibliography

The London Underground, an illustrated history, by Oliver Green (Ian Allan Ltd, 1987).

Moving Millions, a pictorial history of London Transport, by Theo Barker (London Transport Museum and Book Production Consultants 1990).

London Underground Architecture, by David Lawrence (Capital Transport Publishing 1994).

Bright Underground Spaces – the Railway Stations of Charles Holden, by David Lawrence (Capital Transport Publishing 2008).

The Jubilee Line Extension, by Kenneth Powell (Laurence King Publishing 2000).

Rails Through The Clay, by Alan Jackson and Desmond Croome (Capital Transport Publishing 1993).

Underground – how the Tube shaped London, by David Bownes, Oliver Green and Sam Mullins (Allen Lane, an imprint of Penguin Books 2012).

The Subterranean Railway – How the London Underground was built and how it changed the city for ever, by Christian Wolmar (Atlantic 2004).

Index